T0146385

Car Design

Hans-Ulrich von Mende

Car Design
Von der Kutsche zur Auto-Mobilität
From the Carriage to Auto-Mobility

Edition Axel Menges

Für Elinor, Ada, Leo, Luis, Anna, Tyll, Maike und
Eva, die mir Zeit gaben für das Buch.
For Elinor, Ada, Leo, Luis, Anna, Tyll, Maike, and
Eva, who gave me time for this book.

© 2018 Edition Axel Menges, Stuttgart/London
ISBN 978-3-86905-010-2

Druck und Bindearbeiten / Printing and binding:
Graspo CZ, a.s., Zlín, Tschechische Republik /
Czech Republic

Übersetzung ins Englische / Translation into English:
Ilze Mueller
Design: Axel Menges
Layout: Helga Danz

6	Vorwort
8	1. Ursprünge. Die Technik bestimmt die Form
14	2. Von 1900 bis 1949. Die Technik arrangiert sich mit der Form
24	3. Von 1950 bis heute. Die Form bestimmt die Technik
36	4. Verrückte Technik schafft skurrile Form
44	5. Form und Technik machen sich klein
48	6. Form und Technik sind Aufgaben für Designer
56	6.1. Form und Technik schaffen Bilder
62	6.2. Form und Technik bilden ein Konzept
72	6.3. Form und Technik passen sich dem Wind an
78	6.4. Die Form hilft der Technik bei der Sicherheit
82	7. Form und Technik formen das Gesicht eines Autos
86	7.1. Die Form wird kopiert, die Technik nicht
90	7.2. Die Technik schenkt der Form die Flanken
96	7.3. Keine Technik, viel Form bei schnellen Linien
100	7.4. Form als Keil, Technik angepaßt
102	7.5. Die Technik öffnet sich für die Form
110	7.6. Technik und Form erlauben negative Heckfenster
114	7.7. Die Technik verschieden, die Form ein Vorbild aus den USA
116	7.8. Die Technik freut sich über die Form
122	7.9. Form und Technik glänzen am Heck
126	7.10. Technik und Form in Schwarz zur optischen Täuschung
132	7.11. Die Form braucht Farbe, die Technik nicht
136	7.12. Form und Technik müssen sich einig sein
142	Nachwort
144	Glossar
150	Bibliographie

7	Preface
9	1. Origins. The technology determines the form
15	2. From 1900 until 1949. Technology comes to terms with form
25	3. From 1950 until today. The form determines the technology
37	4. Crazy technology leads to bizarre form
45	5. Form and technology go small
49	6. Form and technology are up to designers
57	6.1. Form and technology create images
63	6.2. Form and technology constitute a concept
73	6.3. Form and technology adapt to the wind
79	6.4. The form helps technology improve safety
83	7. Form and technology shape a car's look
87	7.1. The form is copied, though not the technology
91	7.2. Technology determines the form of the flanks
97	7.3. No technology, plenty of form with fast lines
101	7.4. Form as a wedge, technology follows suit
103	7.5. Technology opens to let in form
111	7.6. Technology and form make negative rear windows possible
115	7.7. The technology is different, the form is modeled on an American one
117	7.8. Technology delights in form
123	7.9. Form and technology gleam on the rear
127	7.10. Technology and form in black create an optical illusion
133	7.11. Form needs color, technology does not
137	7.12. Form and technology must be in agreement
143	Afterword
145	Glossary
151	Bibliography

Vorwort

Geschwindigkeit war und ist faszinierend – egal wie man sie erlebt: zu Fuß, auf Skiern, auf dem Pferderücken, auf dem Fahrrad, motorgetrieben auf zwei, drei oder vier Rädern. Alles, was außer Muskelkraft hinaus beschleunigt, nutzt technische Entwicklungen. Den größten Sieg hat das Auto errungen. Wenn auch Geschwindigkeitsbegrenzung weltweit der Standard ist, dem leistungsgedopten Autos tat und tut es keinen Abbruch. Stößt der Explosionsmotor wegen seiner Abgase an Grenzen, so hat der elektrische Antrieb noch viele Entwicklungsjahre vor sich, aber in Sachen Tempo ist er dem Verbrennungsmotor gleichwertig und zunehmend im Vorteil. Es bleibt das Faszinosum der Geschwindigkeit. Fahrzeuge, die sich leistungsmäßig nicht profilieren können, suchen den Ausweg in der optischen Aufrüstung. Automobiles Design, zu Beginn bescheidenes Beiwerk bei der Formfindung, wird zum Maskenbildner emotionsbetonter Karosserien. Funktionen wie Aerodynamik als wesentliche Hilfe, geschwindigkeitsfördernd zu wirken, verbergen sich in raffinierten Details.

Was sich in der äußeren Form manifestiert, wird konsequent und sinngemäß im Wageninneren fortgeführt. Der Fahrerplatz mit allen Anzeigen nennt sich nun Cockpit. Gepolsterte Fenstersäulen, Airbags und weiche Oberflächen sind die Beruhigungspillen für gefühlte Geborgenheit. Grundversionen eines Modells werden mit farbigen Akzenten, vorzugsweise in Rot, in leistungsgesteigerten Varianten optisch unterscheidbar gemacht. Trotz vorgegebener, in Zukunft einzuhaltender Abgaswerte bringen speziell PS-starke Marken Autos heraus, die an Stärke und Größe zunehmen. Dagegen gibt es die verständliche Zurückhaltung der Kunden beim Kauf von Elektroautos. Reichweite, Batteriekapazität, Ladezeiten und fehlende Infrastruktur von Ladestationen können die aktuellen Abgaswerte noch nicht kompensieren. So müssen Automobildesigner die gegensätzliche Aufgabenstellung bewältigen: einerseits das klassische Automobil einkleiden, andererseits das reine Elektromobil vorstellen. Neben den ersten Serienmodellen mit E-Antrieb werden sogenannte Concept-Cars gezeigt, die neue Formen für den neuen Antrieb suchen.

Das Buch zeigt neben dem historischen Weg der Karosserienentwicklung die Detailarbeit im Design. Wenn sich in den letzten Jahrzehnten die grundsätzlichen Arbeitsschritte auch nicht wesentlich verändert haben, so können spezielle elektronische Hilfsmittel der photographischen Darstellung und des Modellbaus die Entwicklungsstufen eines Modells extrem beschleunigen. Dabei ist festzustellen, daß die originäre Qualifikation eines Designers nicht vernachlässigt werden darf: der Umgang mit Stift und einem Blatt Papier. Die anerkannte Arbeit des Designers hat ihn in der Hierarchie eines Konzerns aufsteigen lassen, leider aber noch zu selten im Range eines Vorstands. Tröstlich dabei ist, daß seine Arbeitsbedingungen verbessert, daß seine Studios in elegante Säle verwandelt oder an internationale Plätze verlegt wurden, um der Inspiration auf die Räder zu helfen. Passend dazu mutierte die Ausbildung zum Automobildesigner zum harten Training, bei dem Kreativität fast wichtiger erscheint als Technikverständnis, geschuldet dem Hunger nach differenzierter Formgebung des automobilen Prinzips. Diese Entwicklung hat sich von der einstmals verfolgten Maxime der funktionsgerechten Gestaltung abgekoppelt. Kann es sein, daß der Druck nach Originalität auch den häufigen Wechsel von leitenden Automobildesignern fördert? Waghalsig ist die Vorstellung, daß diese Wissensträger die Kenntnis der kommenden Modelle des alten Arbeitgebers bei dem neuen Dienstherren kaum ausschalten können. Dies steht im Gegensatz zum Bemühen um strikte Geheimhaltung neuer Modelle. Parallel dazu beteiligt sich die Fachpresse daran, dank perfekter Computerprogramme photorealistische Abbildungen mit der Behauptung zu zeigen, so sehe das kommende Modell XY aus, und das alles im Ergebnis gern mit modischen Tendenzen. Vieles verliert so an Einzigartigkeit, und Beliebigkeit wird zum Gift des Markenbilds.

Dieses Buch schärft mit vielen Abbildungen das Auge, denn einerseits prägt die technische Entwicklung die Form des Automobils, andererseits machen aktuelle Beispiele formaler Details Unterschiede klar, die Marken stabilisierendes Design ausmachen anstatt modischer Trends. Wer aber hat das letzte Wort bei der Produktionsfreigabe: der Finanzvorstand, der Produktionsvorstand, der Marketingvorstand oder der Designer? Für die Hersteller ist alles vermutlich auf einvernehmliche Weise erfolgt. Das erklärt aber nicht, daß jedes neue Modell zum Erfolg wird. Zu sehen ist dies in der jahrzehntealten Liste der von Fachjournalisten gewählten Autos des Jahres. Als Ausnahme ist ein Modell zu nennen, das dank vernünftiger Technik, zuverlässiger Qualität und zurückhaltendem Design erfolgreicher operiert als Klassenbeteiligte. Gemeint ist der Golf.

Abschließend danke ich dem Karosseriebau Riegelhof für wertvolle technische Informationen. Zu danken ist auch den Herstellern, die bei der Bildbeschaffung großzügig agierten und der Zielsetzung des Buches damit geholfen haben.

Preface

People have always been fascinated by speed – no matter how they experienced it: on foot, on skis, on horseback, on a bicycle, motor-driven on two, three or four wheels. Every method of acceleration except for muscular strength uses technical developments. Foremost among these is the car. Even though speed limits are the rule all over the world, cars, high on speed, have never been affected by this fact. While the internal combustion engine has its limitations because of its exhaust gases, the electric drive still has many years of development ahead, but is equivalent to the combustion engine as far as speed is concerned. Speed continues to fascinate us. Vehicles that can't make their mark in terms of performance make up for it in terms of visual upgrading. Car design, initially a modest appendage in finding a form for the automobile, now has the role of creating emotionally charged bodywork. Functions such as aerodynamics, a major method of promoting speed, are concealed behind stylish details.

Trends expressed in the exterior form of the car are consistently continued in its interior. The driver's compartment with all its displays and controls is now called a cockpit. Upholstered window pillars, air bags and soft panels create a sense of calm and security. The basic versions of a model are visually differentiated by means of colorful accents, preferably red, in performance-enhanced variants. Despite prescribed emission values that must be adhered to in the future, certain high-performance brands have been manufacturing increasingly powerful and large cars. On the other hand, consumers have been understandably reluctant to buy electric cars. Range, battery capacity, charging times and lack of charging station infrastructure cannot as yet make up for current emission values. That is why car designers have to solve two contradictory problems: creating the bodywork for the classic car on the one hand while introducing the all-electric car on the other hand. In addition to the first production models with electrical drive they are showing so-called concept cars that seek new forms for the new drive.

In addition to telling the history of how the automobile body developed, the present book describes the evolution of design-related details. Even though the basic steps of design work have not substantially changed in recent decades, specific electronic tools such as photographic reproduction and model construction can greatly accelerate the stages of development of a model. At the same time the primary skills of a designer must not be neglected: working with a pencil and a sheet of paper. The fact that the work of designers is openly acknowledged means a rise in the hierarchy of the corporation for them; unfortunately they have too rarely risen to the rank of executives. The good news is that their working conditions have improved, and that their studios have now been transformed into elegant halls or moved to international venues in order to encourage the flow of inspiration. Appropriately, the training of an automobile designer has become a tough workout in which creativity almost appears to be more important than technical expertise, owing to the strong pressure to make the design of every model different. This development no longer pursues the maxim once observed by designers – functional design. Is it possible that the pressure to be original also promotes the frequent changeover of leading car designers? It is daunting to think that the designers can hardly forget what they know about the coming models of their former boss when they start working for their new employer. This runs contrary to the strict secrecy that surrounds new models. At the same time, thanks to perfect computer programs, the trade press is involved in publishing realistic pictures that claim to show what the coming model XY looks like, preferably including all the trendy details. This means a loss of distinctiveness, and the arbitrary nature of designs proves toxic for the brand image.

The many illustrations found in this book will sharpen our eyes, for on the one hand technical development determines the form of a car, while on the other hand current examples of formal details make clear the differences between a design that stabilizes brands instead of stylish trends. But who has the last word regarding production release: the chief financial officer, the production manager, the head of marketing or the designer? As the manufacturers are concerned, it was all a consensus-based process. But that doesn't mean that every new model will become a success. We can see this from the decades-old list of Cars of the Year selected by specialist journalists. Only the exceptional model – thanks to sound technology, reliable quality and restrained design – functions more successfully than its competitors. The car I mean is the VW Golf.

In conclusion I would like to thank Riegelhof Karosseriebau for valuable technical information. My thanks also go to the manufacturers who generously shared pictures and thus helped the book to achieve its goal.

1. Ursprünge. Die Technik bestimmt die Form

Vielleicht war es ja so: Unser Urahn hatte gerade eine Reisepause gemacht, saß am abschüssigen Ufer, sein Pferd graste hinter ihm, befreit von der schweren Astgabel, die die Habseligkeiten seines Halters aufnahm und wie ein Schlitten gezogen werden mußte. Das Ganze muß mehr als 5000 Jahre her gewesen sein, als dann der Urahn mit einem runden Kiesel spielte und ihn den Hang hinunterrollen ließ. War das die Erfindung des Rades? Eher nicht, aber ein Anfang, denn unser Urahn mußte noch auf die Idee eines zentralen Loches und einer Achse kommen, um es Rad nennen zu können. Als das geschafft war, machte er sich Gedanken, wie Rad und Trägergerüst zu verbinden seien. Da Faulheit die Mutter der Erfindung ist, kam er bald auf die Idee dieser Verbindung und hatte somit seinen einachsigen Wagen, der ihn schneller als mit dem Schlitten voranbrachte.

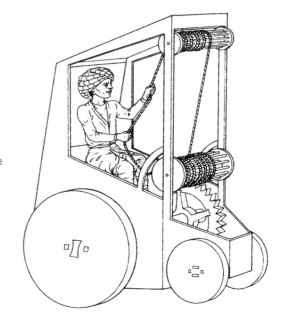

Erste Funde zweiachsiger Wagen stammen aus dem Zweistromland, geschätzt auf 5000 Jahre vor unserer Zeit. War ein einachsiger Wagen leicht zu lenken, so mußte der zweiachsige einen Weg für die Vorderachse finden, um Kurven zu nehmen. Einfach ging es bei kurzem Radstand mit seitlichem Druck auf die Deichsel. Je länger diese, desto geringer ist der Kraftaufwand dank des Hebelgesetzes. Die bessere Lösung kam mit der Drehschemellenkung auf. Hier schwenkte die vordere Achse um ihren Mittelpunkt, ein Prinzip, das bis heute speziell bei Anhängern überlebt hat. Die Römer waren Meister im Bau ihrer »Carruca« (daher stammt unser Wort »Karre«), des geschlossenen Reisewagens, ein bis heute gültiges Prinzip. Noch waren Vorder- und Hinterräder gleich groß, und daher gelang der Einschlag der Vorderachse nur so weit, wie die Räder dem Aufbau nahekamen. Erst der sogenannte Durchlauf bei Kutschen und kleinere Vorderräder konnten den Lenkeinschlag auf 90 Grad erhöhen. Ein Nachteil dieser Verbesserung: Bildeten die vier Räder im Geradeauslauf noch vier Aufstandflächen, so reduzierte sich das bei dem 90-Grad-Manöver auf prinzipiell drei, weil die Vorderradachse die Verlängerung des Achsmittelpunkts der Hinterachse bildete.

Die Römer erkannten trotz ihres gepflasterten Straßennetzes (z. B. der Via Appia), daß große Räder kleine Unebenheiten besser überwanden, aber Stöße weiterhin ungefedert in den Aufbau übertrugen. Die Lösung war die Aufhängung des Aufbaus mittels Lederriemen, eingehängt an U-förmige Rahmen. So ließ sich die Transportleistung in Sachen Tempo im Vergleich zur Schrittgeschwindigkeit verdreifachen. Dieses Prinzip wurde mit den Jahrhunderten verfeinert. Der ungarische Ort Kocs bescherte uns im 15. Jahrhundert nicht nur eine verbesserte Federung, sondern war auch der Ursprung für unser Wort »Kutsche«. Zu jener Zeit gab es auch erste Bemühungen, die Zugkraft des Vierbeiners zu ersetzen. Versuche dieser Art unterschieden sich prinzipiell durch zwei Antriebsarten. Die eine zog den Wagen, die andere wurde durch »bordeigene« Kraftübertragung auf die Räder bewegt. So war bei Version eins der Wind der Kraftspender, der mit bordeigenen Segeln oder Luftpaddeln Vortrieb bringen sollte. Version zwei versuchte es mit Maschinen, wie bei Leonardo da Vinci und seinem durch Federn gespannten Antrieb, einem Uhrwerk nicht unähnlich. Version zwei mußte aber kläglich scheitern, weil die Kraftübertragung viel zu schwach war, ein entsprechendes Drehmoment zum Drehen der Räder bereitzustellen. Hinzu kamen fehlende glattflächige Straßen, die Radbewegungen wesentlich erleichtern können.

Erst die Erfindung der Dampfmaschine, später die Explosionsmotoren und noch später der Elektroantrieb machten den Traum vom Automobil, dem »Sich-selbst-Bewegenden«, wahr. Krachte der Franzose Nicolas Cugnot 1769 mit seinem Dampf-Ungetüm, dreirädrig und mit Frontantrieb, noch gegen eine Mauer, so konnten bereits 1821 dampfbetriebene Busse den Liniendienst in England aufnehmen. Das Auto im Wortsinne war also längst geschaffen, die Geschichtsschreibung bevorzugt aber erst das 1886 an Carl Benz erteilte Patent als Geburtsstunde des Automobils. Vorher gab es bereits mehrere Versuche mit Explosionsmotoren, alle mit sogenannter innerer Verbrennung, im Gegensatz zum Dampfantrieb mit äußerer. Erfindungen wie der Motorwagen von Siegfried Marcus 1870 in Österreich oder der gasbetriebene Motor 1860 von Etienne Lenoir in Paris blieben Einzelversuche. Die Erfindung von Nicolaus Otto 1862, eines Viertaktmotors, und später der Dieselmotor von Rudolf Diesel aus dem Jahre 1893, der sowohl zwei- als auch viertaktig laufen kann, eroberten die pferdelosen Kutschen als Antriebsersatz.

Wer aber das Pferd ersetzen will, muß sich Gedanken machen, wie das neue Gefährt gebremst, gelenkt und angetrieben wird. Zunächst waren Bremse und Lenkung noch von Kutschen geprägt: Klotzbremse, auf die Radreifen wirkend, Drehschemellenkung und Riemenantrieb waren Ansätze. Der Aufbau, kutschengleich, also eher offen als geschlossen und, wenn offen, dann mit Faltdach, hatte noch nicht das uns bekannte Bild des Automobils. Zur besseren Manövrierung trug die neue Erfindung der Achsschenkellenkung bei. Erstmalig 1761 von einem gewissen Erasmus Darwin erfunden, dann 1816 vom Münchner Stellmacher Georg Lankensperger zum Patent

1. 1420 versuchte Giovanni da Fontana, mit Muskelkraft und einer Art Flaschenzug – in der Wirkung wie ein einfaches Getriebe – voranzukommen. Sollte es funktioniert haben, war es sicher ein langsames Reisen. (Photo: Beatsrabban.)
2. 2000 Jahre war es her, als die Römer so auf Reisen gingen. Bis heute hat sich kaum etwas im Kutschbau geändert – und die Grundform für das »Selbstbewegende«, das Automobil – ist zu erahnen. (Photo: N. von Kospoth.)
3. 1495 hat Leonardo da Vinci mit einem Federwerk als Antrieb das »Selbstbewegende« erdacht. Ein Nachbau bewies die Funktionsfähigkeit, aber auch die fehlende Kraft, etwa einen Menschen zu transportieren. (Photo: Mechsouvenir.)
4. 1649 machte sich der Holländer Simon Stevin den Wind zur Antriebskraft. Stark genug für ein Schiff auf Rädern, aber hilflos bei Windstille und fehlender glatter Straßen. (Stich von Jacques de Gheyn.)

1. In 1420, Giovanni da Fontana tried to move from place to place by means of muscle power and a kind of block and tackle – which worked like a simple transmission. If it really did function, it was undoubtedly a slow way of traveling. (Photo: Beatsrabban.)

2. Two thousand years ago, this is how the Romans traveled. To this day hardly anything has changed in the construction of carriages – and there's already a hint here of the basic form of the »self-moving« vehicle, the automobile. (Photo: N. von Kospoth.)

3. Leonardo da Vinci, in 1495, invented the »self-moving« vehicle, which had a spring drive. A replica has demonstrated that it is capable of functioning, but also that it lacks the power to transport, say, a human being. (Photo: Mechsouvenir.)

4. In 1649 the Dutchman Simon Stevin made the wind his propelling force: strong enough for a ship on wheels, but helpless during a calm and in the absence of smooth roads. (Engraving by Jacques de Gheyn.)

1. Origins. The technology determines the form

Maybe this is how it happened: Our ancestor had just stopped for a rest, and sat down on a steep bank. His horse was grazing behind him, free of the heavy forked branch that was laden with the belongings of its owner and had to be dragged like a sled. It must have been more than 5000 years ago that our forefather began playing with a round pebble and let it roll down the slope. Was that the invention of the wheel? Probably not, but it was a start, for our forefather still had to invent a central hole and an axle before he could call it a wheel. Once he had managed to do that, he wondered how the wheel and the carrier frame could be connected. Since idleness is the mother of invention, he soon hit on the idea of this connection and thus had his single-axle cart, which moved him forward faster than when he used the sled.

The first two-axle carts were found in Mesopotamia, an estimated 5000 years before our era. While a single-axle cart was easy to steer, a way had to be found for the front axle of the two-axle cart to take curves. When the wheelbase was short, this was simple, with lateral pressure on the shaft. The longer the shaft, the less energy it took, thanks to the principle of the lever. The better solution came with turntable steering. Here the front axle pivoted around its center, a principle that has survived to this day particularly in trailers. The Romans were masters at building their »carruca« (hence our word »car«), the closed carriage, a principle that is still valid today. At the time the front and rear wheels were still the same size, which is why the steering angle of the front axle was only successful as long as the wheels were close to the carriage body. It was only when carriages had so-called cutouts and smaller front wheels that the steering angle could be increased to 90 degrees. This improvement had one disadvantage: While the four wheels still formed four areas of contact when the carriage moved in a straight line, these were basically reduced to three when it made a 90-degree turn, because the front wheel axle was the extension of the rear axle's center point.

Though they had a network of paved roads (e. g., the Via Appia), the Romans realized that large wheels were better at negotiating small bumps, while – having no springs – they continued to transfer shocks to the carriage. The solution was to suspend the body of the carriage by means of leather straps from a U-shaped frame. In terms of speed, transport capacity as compared to walking speed was tripled. This principle was improved over the centuries. In the 15th century, the Hungarian village of Kocs not only gave us an improved suspension, but was also the origin of our word »coach«. That era also saw the first efforts to replace the traction force of

angemeldet, weitergereicht an den Engländer Rudolph Ackermann und schließlich 1876 verfeinert durch Amédée Bollées Patent und 1891 durch das Patent von Carl Benz, ist die Achsschenkellenkung bis heute weltweit im Automobilbau die (fast) einzige Lenktechnik. Hier wird jedes Vorderrad einzeln eingeschlagen. Der größte Vorteil: Wegen der Drehschemellenkung hochgebaute Fahrzeuge oder Fahrzeuge mit dem sogenannten Durchlauf konnten abgelöst werden durch die neue Lenktechnik, die für Pferdekutschen wenig geeignet, für motorangetriebene Gefährte aber nahezu ideal war. Warum? Weil die Bauhöhe des Fahrzeugs erheblich reduziert werden konnte und der Wagenkörper an den Vorderrädern lediglich etwas schmäler ausgebildet werden mußte, um dem eingeschlagenen Rad Platz zu lassen.

Hatten inzwischen eingeführte Blattfedern bei Kutschen und deshalb auch bei ersten Automobilen den Komfort erhöht, so mußten sich die Räder zunächst mit der klassischen Holzkonstruktion aus Radreifen, Speichen und Radlager begnügen. Zudem war die Aufstandsfläche sehr klein wegen der entsprechend schmalen Radreifen, die aus Haltbarkeitsgründen mit Flachstahlbändern umgürtet waren. So war ein weiteres Mal nicht die Faulheit, in diesem Fall aber die Bequemlichkeit die Mutter der Erfindung. Kleinere Unebenheiten lassen sich nämlich durch ein Polster abfedern. Wenn das Polster luftgefüllt ist und sich um das Rad legen kann, ist die Erfindung des Luftreifens geglückt. Das gelang patentreif 1845 dem Schotten William Thomson aus Edinburgh. 1888 schob der Arzt John Boyd Dunlop eine verbesserte und patentierte Version mit Ventil nach. André Michelin aus Frankreich dachte weiter und kam 1891 auf die Idee des demontierbaren Reifens. Die waren im übrigen zunächst hellbeige bis weiß wegen des Kautschuk-Basismaterials. Erst das spätere Beimischen von Ruß verbesserte die Haltbarkeit und raubte dem Rad wegen des schwarzen Reifens eine gewisse Eleganz. Dies änderte sich wieder mit dem Aufkommen der Weißwandreifen Anfang der 1930er Jahre.

Das alles war Carl Benz nicht wichtig, als er sein Automobil 1886 zum Patent anmelden konnte. Es war eher ein Fahrrad mit Stützrädern als der optische Vorbote des Automobils. Drei sehr zierlich bespeichte Räder mit Hartgummibereifung, eine ledergepolsterte Sitzbank und ein darunter geschobener, flach liegender Ein-Zylinder-Ottomotor boten optisch schiere Technik. Das kleinere Vorderrad erleichterte etwas den Kraftaufwand beim Lenken, die schmalen Räder waren höchst ungeeignet für unbefestigte Straßen: Aber es war das erste Auto, so man der Geschichtsschreibung folgen will. Diese täuschte sich auch in der ersten Darstellung der Ausfahrt von Bertha Benz mit ihren Söhnen von Mannheim nach Pforzheim und zurück im Jahre 1888. Viele nachgestellte Bilder zeigen sie mit dem Gefährt von 1886. Tatsächlich war es die automobile Variante mit drei starken Holzspeichenrädern und erstarktem Holzaufbau. Parallel zu Benz hatte Gottfried Daimler 1886 auch ein Automobil zu bieten. Das hatte vier Räder, war praktisch eine offene Kutsche ohne Pferde und verließ sich auf bewährte Bauart bei verbessertem Motorantrieb, was aber für ein Patent nicht reichte.

Die Geburt des Automobils ließ seinen Sprößling schnell erstarken. Unzählige erste Versuche erhofften sich gute Geschäfte. Besonders Frankreich hatte schnell das automobile Prinzip als

5. 1769 war die Dampfmaschine der Motor für das Vorderrad. Technisch richtig gedacht von Nicholas Cugnot, formal ohne Zukunft. (Photo: Thesupermat.)

6. 1875 versuchte sich Siegfried Marcus an einem motorgetriebenen Gerät, das als Basis ein Kutschenwagen war. (Photo: P. Diem.)

7. 1884, zwei Jahre vor dem Benz-Patent, glaubte Enrico Bernardi die Lösung gefunden zu haben mit einer Minimalstlösung: stehender Zylinder nutzte den Riemen als Kraftübertragung. (Photo: D. Strohl.)

8. 1833 war der Dampfantrieb so zuverlässig, daß Linienbusse damit die Stationen bedienen konnten. Zum Beispiel der Enterprise Steambus von Walter Hancock, hier als Nachbau. (Photo: Wikipedia.)

5. In 1769 the steam engine was the motor for the front wheel. Technically correctly designed by Nicholas Cugnot, but formally without a future. (Photo: Thesupermat.)
6. Siegfried Marcus, in 1875, had a go at a motor-driven vehicle based on a coach. (Photo: P. Diem.)
7. In 1884, two years before the Benz patent, Enrico Bernardi believed he had found the answer with a minimal solution: A vertical cylinder used a belt for power transmission. (Photo: D. Strohl.)
8. Steam power was so reliable in 1833 that public buses could serve the stations – for example, Walter Hancock's Enterprise steam bus. Pictured here is a replica. (Photo: Wikipedia.)

draft animals. Experiments of this kind fall into two categories depending on the type of drive. One pulled the carriage, the other was propelled by the »on-board« transmission of power to the wheels. Thus in version one the wind provided the power: On-board sails or air paddles were supposed to propel the vehicle. Version two tried the same thing with machines – for instance, Leonardo da Vinci's spring-tensioned drive mechanism, not unlike clockwork. But version two was doomed to fail miserably because the power transmission was far too weak to create enough torque to turn the wheels. Moreover, there were no smooth-surfaced roads to make the turning of the wheels substantially easier.

Only the invention of the steam engine, followed by internal combustion engines and later the electric drive, made the dream of the automobile, the vehicle that »moved itself«, come true. While the French engineer Nicolas Cugnot and his 1769 steam monster, three-wheeled and with a front-wheel drive, still ran up against a wall, as early as in 1821 steam-driven buses were able to begin a regular passenger service in England. Thus the car in the true sense of the word had been created long ago, though historians prefer to designate only the patent issued to Carl Benz in 1886 as the birth date of the automobile. There had already been several previous experiments with engines powered by so-called internal combustion motors, in contrast with steam-driven vehicles that had external combustion engines. Inventions such as the 1870 motorcar of Siegfried Marcus in Austria or the 1860 gas-powered motor of Etienne Lenoir in Paris remained isolated experiments. Nicolaus Otto's 1862 invention, a four-stroke engine, and later the Diesel engine of Rudolf Diesel in 1893, which is designed as either a two- or four-stroke cycle, were used in horse-less carriages everywhere as an alternative power source.

But if one wants to replace the horse, one must figure out how the new vehicle is to be braked, steered and powered. Initially the brake and steering were still influenced by coaches: brake shoes applied to the tires, turntable steering and a belt drive were early concepts. The body, which was similar to that of a coach, i.e., open rather than closed and, if open, had a folding roof, did not yet look like the car we now know. The newly invented axle steering device helped improve maneuvering. First invented in 1761 by a certain Erasmus Darwin, then patented in 1816 by the Munich wheelwright Georg Lankensperger, passed on to the English inventor Rudolph Ackermann and finally improved in 1876 by Amédée Bollée's patent and, in 1891, by the patent of Carl Benz, axle steering has been the (almost) only steering mechanism used worldwide to this day in automobile manufacture. Here every front wheel is turned individually. The biggest advantage: Vehicles that are built high because of turntable steering or vehicles with so-called cutouts could be replaced by the new steering control mechanism, which was not very suitable for horse-drawn carriages, but almost perfect for motorized vehicles. Why? Because the height of the car body could be significantly reduced and the body merely had to be somewhat narrower in shape near the front wheels to leave room for the steered wheel.

While leaf springs had in the meantime been introduced in carriages and therefore increased comfort in the first automobiles as well, the wheels – tires, spokes and wheel bearings – were initially built of wood, according to time-honored tradition. Moreover, their footprint was minimal because of the correspondingly narrow tires that were encircled with flat steel bands to make them more durable. In this case, the mother of invention was not idleness but comfort. The impact of smaller bumps in the road can be absorbed by a cushion. Once the cushion is filled with air and

möglichen Erfolg erkannt, zu nennen wäre da die Marke Panhard & Levassor oder Peugeot. Als Antrieb wurde meist der Explosionsmotor bevorzugt. Aber bereits die Erfindung des Elektromotors machte ihn auch für das Automobil interessant. Offenbar erkannte man schon damals die sich schnell entwickelnde Kraft beim Anfahren und schließlich auch den abgasfreien Betrieb des Motors. Wie schnell es mit der Elektrizität gehen kann, demonstrierte der Belgier Camille Jenatzy mit seinem Gefährt »Jamais-Contente« (die ewig Unzufriedene) 1899. 105,88 km/h las man von der Stoppuhr. Daß es an der speziellen Karosserieform lag, ist zu bezweifeln, denn die zigarrenförmige Karosse hatte frei stehende Räder, und der Pilot Jenatzy ragte ab der Hüfte hoch aus dem Gehäuse – immerhin ist aber der gedankliche Ansatz zu bejahen, daß schnelle Gefährte dem Wind wenig Widerstand bieten sollten.

Das klassische Profil eines Autos mußte aber noch warten. Ansätze dazu waren zunächst gleich große Räder wie bei dem englischen Lancaster von 1895 oder dem Wartburgwagen von 1898 mit einer Vis-à-vis-Sitzanordnung, will heißen, Fahrer und Fahrgäste schauten sich an. Wesentlich mehr »Auto« zeigte 1898 der Daimler 8 PS Phaeton mit Vier-Zylinder-Motor im Bug und Kotflügeln; Bauweise und Profil sind ein noch heute gültiges Prinzip. Waren hier die Räder fast gleich groß, der Motor unverkleidet mit Schlangenrohrkühlung im Bug, so bot 1899 der Renault Type B ein verwirrendes Bild. Zu kurzer Radstand und eine zu hohe verglaste Karosserie – damals noch ganz selten – machten daraus einen Kleiderschrank auf Rädern.

9. 1898 sah das schon aus wie ein Auto – noch vor dem Mercedes 35 PS von 1900: der Daimler 8 PS-Phaeton mit dem ersten Vierzylindermotor. (Photo: Daimler.)

10. Dieses Dreirad reichte 1886 aus, um ein Patent als Auto zu erhalten, was dem Erfinder Benz gefallen hat, aber bei dem wir eher an ein dreirädriges Fahrrad denken. (Photo: Daimler.)

11. 1886 bot Gottlieb Daimler mit seinem Motorwagen vier Personen Platz. (Photo: Daimler.)

12. Der Fiat 3,5 HP hatte bereits 1899 gleich große Räder. (Photo: Fiat.)

13. 1899 konterkarierte der Renault Type B das Thema Tempo, bot dafür Platz für Herren mit Zylinderhut. (Photo: Supercars.)

14. Die Ehrenrunde vor dem Buckingham Palace, heute wie vor 120 Jahren. (Photo: Nicoretro.)

9. In 1898 it already looked like a car – even before the 1900 Mercedes 35 hp: the Daimler 8 hp Phaeton with the first four-cylinder engine. (Photo: Daimler.)

10. In 1886, this three-wheeler was sufficient in order to be patented as a car. Its inventor, Benz, was pleased, while we tend to associate it with a three-wheel bicycle. (Photo: Daimler.)

11. Gottlieb Daimler's motorcar, in 1886, had room for four persons. (Photo: Daimler.)

12. The 1899 Fiat 3.5 hp already had equal-sized wheels. (Photo: Fiat.)

13. In 1899 the Renault Type B was anything but fast, but did have room for gentlemen who wore top hats. (Photo: Supercars.)

14. Trooping the color in front of Buckingham Palace, today as 120 years ago. (Photo: Nicoretro.)

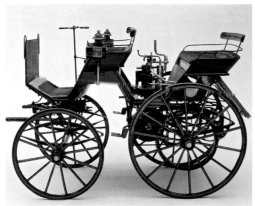

can be placed around the wheel, a pneumatic tire has successfully been invented. A patentable version was successfully developed in 1845 by the Scotsman Robert William Thomson of Edinburgh. In 1888 the physician John Boyd Dunlop followed this up with an improved and patented version that had a valve. The Frenchman André Michelin thought ahead and in 1891 came up with the idea of a removable tire. Incidentally, these tires were initially light beige to white because of the rubber they were made of. A later admixture of soot improved their service life, though the wheels lost a certain elegance due to the black tires. This changed again with the advent of whitewall tires in the early 1930s.

None of this mattered to Carl Benz when he was able to patent his automobile in 1886. It was more a bicycle with stabilizing wheels than the visual precursor of the automobile. Three very gracefully spoked wheels with hard rubber tires, a leather-upholstered bench and a single-cylinder Otto engine placed flat underneath it provided bare-bone technology. The smaller front wheel reduced the amount of strength needed for steering somewhat, and the narrow wheels were highly unsuitable for unpaved roads: But it was the first car, if we want to believe the historians. The latter were also mistaken in their first account of an excursion by Bertha Benz and her sons from Mannheim to Pforzheim and back in 1888. Many reenacted pictures show her with the 1886 vehicle. In fact, however, she rode in the automobile variant that had three strong wheels with wooden spokes and a reinforced wooden body. At the same time as Benz, Gottfried Daimler also had an automobile to offer in 1886. It had four wheels, was practically an open carriage without horses and relied on a proven design along with improved motor drive, which, however, was not enough for a patent.

It did not take long for the new invention to prosper. Countless early experiments gave rise to hope for good business deals. The French in particular had quickly realized that automobiles were a potential success, cases in point being Panhard & Levassor or Peugeot. As a drive, the internal combustion engine was usually preferred. But the invention of the electric motor made it, too, of interest for cars. Obviously people recognized even then how quickly it delivered power when the car was started and appreciated its emission-free operation. In 1899 the Belgian Camille Jenatzy with his vehicle »Jamais-Contente« (never satisfied) demonstrated how fast an electric car could travel. The stopwatch read 105.88 km/h (65.79 mph). It is debatable whether this was due to the specific style of the body, for the cigar-shaped limousine had free-standing wheels, and the pilot Jenatzy towered above it from the hips up. At any rate, one thing was certain: Fast vehicles must be designed to offer the wind very little resistance.

The classic profile of a car was still a thing of the future, however. The initial stages of the modern car were equal-sized wheels, as in the British Lancaster (1895) or the 1898 Wartburg car, which had a vis-à-vis seating arrangement, meaning that the driver and passengers sat face to face. The 1898 Daimler 8 hp Phaeton, which had a four-cylinder front-mounted engine and mudguards, looked much more like our idea of a car; its design and profile are based on a principle that is still valid today. While the Phaeton's wheels were all almost the same size, and the engine, unfaired with a coil cooling system, was in the front, the appearance of the 1899 Renault Type B was bewildering. The wheelbase was too short and the glassed-in body – still quite rare at the time – was too high, making this look like a wardrobe on wheels.

2. Von 1900 bis 1949. Die Technik arrangiert sich mit der Form

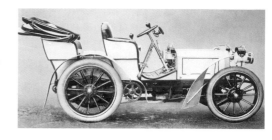

Der Anfang war gemacht. Das Automobil sollte mehr als eine motorisierte Kutsche sein. Schon 1888 führte Preußen die Chauffeurbefähigung ein, quasi den Führerschein. Der Begriff »Chauffeur« ist ein Blick zurück, als der Dampfantrieb üblich war und geheizt werden mußte, was auf französisch eben »chauffer« heißt. Auch vorbei war in England seit 1896 der »Red Flag Act«. Hier verlangte die Obrigkeit, daß eine eine rote Fahne schwingende Person zur Fußgängerwarnung 55 Meter vor einem Automobil zu laufen hatte, was körperlich zumutbar war beim damaligen Tempo der neuartigen Vehikel. Und technisch war »nur« die Achsschenkellenkung schon ausreichend, um das Automobil optisch von der Kutsche zu befreien. Damals ungewohnt, für unser heutiges Auge aber so gewohnt, war zum Beispiel 1900 der Mercedes 35 PS mit fast gleich großen Rädern, einer eindeutig definierten Motorhaube, hintereinander angeordneten Sitzen und Kotflügeln, die sich an der Front vorsichtig von der flächigen, gebogenen Form verabschiedeten, um sich Pflugschaufeln ähnlich an das Fahrgestell zu schmiegen. Lediglich klobige Frontlampen und das fehlende Dach vereinfachten das Bild des Automobils. Im selben Jahr, am 6. März, verstarb Gottfried Daimler, der große Konkurrent von Benz.

Dafür aber wurde ein junger Konstrukteur bekannter: Ferdinand Porsche. Auf der Weltausstellung 1900 in Paris präsentierte die österreichische Firma Lohner sein Auto mit Radnabenmotor, gespeist aus Batterien für die Elektromotoren in den Vorderrädern. Noch war gar nicht ausgemacht, welcher Antrieb das Rennen machen würde: Strom, Dampf oder Benzin. Der Lohner-Porsche machte 50 km/h und schaffte 50 km Reichweite. Allerdings verzichteten die Elektromotoren auf die komplizierte Kraftübertragung auf die Hinterräder – meist mit Kette, Kupplung, Kardanwelle und Differential –, machten aber das Lenken schwerer, weil die ungefederte Masse der Räder größer und schwerer war als übliche Fronträder, denn Servolenkung wurde erst 1926 erfunden und ab 1950 serienmäßig verwendet. Viele Automobile hatten mit ihrer Hecklastigkeit weniger Auflast auf den Vorderrädern und so leichteres Spiel mit der Lenkung, manchmal ein zu leichtes. Ebenfalls in 1900 hatte Porsche auf Kundenwunsch vier Radnabenmotoren gebaut und schuf so das erste allradgetriebene Automobil. Den Nachteil des hohen Gewichts der Batterien kompensierte Porsche 1902 mit dem, was er drei Jahre zuvor geplant hatte: den »Range Extender«, den benzinbefeuerten Reichweitenverlängerer. Formal kam damit die Motorhaube zurück, auf die der erste Lohnerwagen verzichten konnte. Auch zurück kam das ungeliebte Abgas. War es das, was den deutschen Kaiser veranlaßte zu sagen, er setze lieber auf das Pferd als auf das Auto?

Elektroantrieb hatte einen vielfach höheren Wirkungsgrad als Explosionsmotoren. Das bewies ja schon Jenatzy 1899 mit dem Knacken der 100-km/h-Marke. Ungebrochen war aber der Drang nach Tempo, egal ob elektrisch, dampfbefördert oder benzingetrieben. Dagegen standen zunächst das herkömmliche Bauprinzip des Automobils mit Fahrgestell, Starrachsen vorne und hinten, hohen Aufbauten mit und ohne Dach, Gewicht und natürlich das Unverständnis für aerodynamische Grundformen. Das erkannten nur wenige und versuchten sich mit geschlossenen, glatten Grundkörpern, scheibenförmigen Rädern (anstelle von Speichen), leichtem Gewicht, aber im Antrieb sowohl elektrisch, dampfbetrieben als auch mit Benzin. Aus den USA kennen wir den elektrischen Walter Baker Torpedo (einer der aufkommenden »Karosserieformen«) von 1902 und

15. 1901 war der Lohner Porsche höchst modern mit Elektromotoren in den Vorderrädern. Eine Bauart ohne Kardanwelle und Differential. Heute wieder eine Variante mit Elektroantrieb. (Photo: Porsche.)
16. 1900 wagte sich der Mercedes 35 PS noch nicht, gleich große Räder zu wählen. Dafür hatte auch er wie der Fiat von 1899 ein Faltdach als kargen Wetterschutz. (Photo: Daimler.)
17. 1902 schaffte der Walter Baker Torpedo elektrisch über 100 km/h. Scheibenräder und tropfenförmige Karosse halfen dabei. Als Rekordwagen zwar ästhetisch, aber wenig alltagstauglich wie alle sportlich ambitionierten Wagen. (Photo: Rockemneedles.)
18. 1902 hatte der Serpollet Type H seinen Schiffsbug aerodynamisch mit dem davor gelegten Kühler verdorben. Kein Problem, denn Pilot und Beifahrer taten ein übriges, dem Wind die Stirn zu bieten. (Photo: Citedelautomobile.)
19. 1906 zeigte der Renault GP eine Kühlerhaube, die formal Jahrzehnte hielt, und Nachahmer fand, aber auch sofort Erkennungszeichen war. Möglich war es, weil der Kühler hinter dem Frontmotor lag. (Photo: Autokaleidoskop.)

15. The 1901 Lohner Porsche, with its hub-mounted front wheel electric motors, was ultra-modern – a design without a drive shaft and differential. Today, again, a version with electric drive. (Photo: Porsche.)

16. In 1900 the Mercedes 35 hp did not have the guts to opt for equal-sized wheels. On the other hand, like the 1899 Fiat, it did have a retractable roof, barely enough to protect passengers from inclement weather. (Photo: Daimler.)

17. The electrically driven 1902 Walter Baker Torpedo did over 100 km/h. Its disc wheels and drop-shaped body were a help. While aesthetically pleasing as a record car, it was as unsuitable for daily use as all vehicles that aspire to be a sports car. (Photo: Rockem-needles.)

18. In 1902 the Serpollet Type H had aerodynamically spoiled its ship's bow by placing the radiator in front of it. Not a problem, since the pilot and front-seat passenger also had to brave the wind. (Photo: Citedelautomobile.)

19. The 1906 Renault GP had a radiator hood that formally continued to be used for decades and had its imitators, while it also made for immediate brand recognition. What made this possible was that the radiator was located behind the front-mounted engine. (Photo: Autokaleidoskop.)

2. From 1900 to 1949. Technology comes to terms with form

A start had been made. The car was to be more than a motorized carriage. As early as 1888 Prussia introduced the chauffeur's qualification certificate, virtually a driver's license. The term »chauffeur« is a look backward to a time when steam engines were in common use and had to be heated, for which the French word is »chauffer«. Another thing of the past is the British »Red Flag Act« of 1896. It was a policy requiring that a person waving a red flag was to run 55 m ahead of an automobile to warn pedestrians of its approach, which was physically possible, considering the speed of the novel vehicles at the time. And technically the »mere« fact that a car had axle-pivot steering was sufficient to distinguish the car visually from a carriage. Unusual at the time, but so familiar to our eyes today, was the Mercedes 35 hp (1900), to name one example, with almost same-sized wheels, a clearly defined hood, tandem seats and mudguards that in front tentatively departed from the two-dimensional, curved form, hugging the chassis like plow-shares. Only bulky headlights and the fact that there was no roof simplified the appearance of the automobile. In the same year, on March 6th, Gottfried Daimler, Benz's great competitor, died.

Meanwhile, however, one young engineer became better known: Ferdinand Porsche. At the 1900 Paris world's fair the Lohner company of Austria exhibited his car, which had a hub motor, powered by batteries for the electric motors in the front wheels. It was not yet certain which technology would win the race: electricity, steam or gasoline. The Lohner Porsche averaged 50 km/h and managed a range of 50 km. It is true that the electric motors did not have the complicated power transmission to the rear wheels – usually by means of a chain, coupling, drive shaft and differential –, but made steering more difficult because the unsprung mass of the wheels was larger and heavier than standard front wheels, for power steering was not invented until 1926 and was not mass-produced until 1950. Many cars, being tail-heavy, had less load on the front wheels, which made steering easier, and sometimes too easy. Also in 1900, Porsche had built four customized wheel-hub motor models and thus created the first all-wheel-drive car. In 1902 Porsche compensated for the disadvantage of the batteries' heavy weight, something he had planned three years earlier: the »range extender«, which was gasoline-powered. Formally, this marked the return of the hood, which the first Lohner car had been able to dispense with. It also meant the return of undesirable exhaust emission. Was it this that caused the German emperor to say he would rather back the horse than the motorcar?

Electric motors were far more efficient than internal combustion engines. This was proved as early as 1899 by Jenatzy when he broke the 100-km/h limit. But the desire for speed was unbroken, regardless of whether it was electric, steam- or gasoline-driven. Initially it was impeded by the conventional construction of the car, with a chassis, rigid axles in the front and back, high bodies with and without a roof, weight and naturally also poor understanding of basic aerodynamic forms. Only few people realized this and experimented with closed, smooth bodies, disc-shaped wheels (instead of spokes), light weight, but engines that were powered by electricity as well as steam or gasoline. We are familiar with the 1902 electric Walter Baker Torpedo (one of the up-and-coming bodywork shapes) and the 1907 Stanley Steamer (i.e., steam-powered car), from the U.S., and the 1902 Serpollet Type H and the 1906 Renault GP, from France. The Serpollet, with its coil cooling system studded with cooling ribs in front of its pointed nose, had possibly done away with a potential aerodynamic advantage. The Renault, whose cooling system was behind the motor, had the advantage that its metal hood was slanted in front and only gradually rose toward the back, which reduced drag and ensured that for decades Renault had an eye-catching design of a type only seldom encountered in the cars of other manufacturers. This goes to show that in those days the influence of technical development on the series was still far more important than today, when we are made to believe that Formula One races still fulfill this function. While Serpollet failed as far as the front of the car was concerned, the invention of the German engine designer and Daimler's partner Wilhelm Maybach was an improvement. His honeycomb radiator with its little hexagonal metal cells had a form so free the cooler systems were often specially designed in order to utilize what little there was that distinguished the brand of limousine. Even the so-called pointed radiator with a radiator front that had a vertical kink could make a small contribution to reducing drag.

In Germany the competitive Prince Heinrich Races between 1908 and 1910 promoted closed limousines. The Torpedo or Phaeton design brought smooth-surfaced undercarriages, and slanted mudguards cautiously made the car's appearance dynamic, while the transfer of controls such as the handbrake and gear shift into the car's interior created the term »interior drive«. The steering wheel also had a few additional rings and small hand gears for steering – a long way removed

den Stanley Steamer (also dampfgetrieben) von 1907, aus Frankreich den Serpollet Type H von 1902 und den Renault GP von 1906. Der Serpollet machte mit dem kühlrippenbespickten Schlangenrohrkühler vor dem spitz zulaufenden Bug einen möglichen aerodynamischen Vorteil zunichte. Der Renault hatte mit der Lage des Kühlers hinter dem Motor den Vorteil, die blecherne Motorhaube an der Front schräg zu formen und nur langsam nach hinten ansteigen zu lassen, was den Luftwiderstand reduzierte und Renault ein jahrzehntelanges Wiedererkennungsmerkmal bescherte, wie es nur selten auch andere Hersteller probierten. Zu dieser Zeit war daher der Einfluß der technischen Entwicklung auf die Serie noch viel wichtiger als heute, wo uns weisgemacht wird, daß Formel-1-Rennen diesen Zweck immer noch erfüllen. Was dem Serpollet am Bug mißglückte, konnte die Erfindung des deutschen Konstrukteurs und Daimlerpartners Wilhelm Maybach besser machen. Sein Bienenwabenkühler mit den kleinen sechseckigen Blechkammern war so frei in der Form, daß die Kühlerumrisse sich oft speziell ausbilden ließen, um das Wenige an markentypischer Ausprägung zu nutzen, was die Karossen zu bieten hatten. Selbst der sogenannte Spitzkühler mit in der Senkrechten geknickter Kühlerfront konnte einen kleinen Beitrag zur Reduzierung des Luftwiderstands leisten.

In Deutschland waren die sportlich ausgelegten Prinz-Heinrich-Fahrten zwischen 1908 und 1910 Beförderer der geschlossenen Karosse. Die Bauart Torpedo oder Phaeton brachte glattflächige Unterbauten, schräg auslaufende Kotflügel dynamisierten vorsichtig das Erscheinungsbild, das Hereinziehen von Bedienungselementen wie Handbremse und Schaltung in das Wageninnere schuf den Begriff des Innnenlenkers. Das Lenkrad hatte auch einige zur Motorsteuerung zusätzliche Ringe und Hebelchen – weit vor unseren heutigen mit Schaltern beladenen Ebenbildern. Das geruchsempfindliche, karbidbetriebene Fahrlicht wurde langsam vom elektrischen abgelöst, was mögliche Brände bei Unfällen vermeiden half, denn ab 1908 wurde dies Standard. Das Auto jener Zeit war Ausdruck des Wohlstands, denn die Preise, geprägt von Handarbeit und schwerem Material, konnten sich nur Wohlhabende leisten. Als Beispiel: Ein Satz Luftreifen kostete 1900 umgerechnet 10 000 Euro. Das zu ändern, hatte sich Henry Ford in den USA zur Aufgabe gemacht. Seine Erkenntnis lautete: Je preiswerter ein Auto, desto mehr davon können verkauft werden. Der Erfolg des Modells Tin-Lizzy – von 1908 bis 1927 – war vorgegeben. Bis dahin war es eine der Sparmaßnahmen, das Auto nur in Schwarz zu verkaufen. Dieser Farbton war gängig, preiswert in der Herstellung, schnell trocknend und bequem für die Zulieferer, denn sie mußten alles nur in Schwarz liefern. Im Antrieb hatte sich inzwischen der Explosionsmotor durchgesetzt. So gab es noch 1900 in den USA 1500 Elektroautos, 751 Dampfautos, aber nur 15 Benziner. Der Dampf verlor wegen seiner umständlichen Betreibung, die Batterie wegen Gewicht und Reichweite. Dagegen holte die Benzinmotorentechnik schnell auf und konnte diese Bauart durchsetzen – allerdings nur bis heute. Denn dem E-Antrieb wird wohl die Zukunft gehören.

Was für das Auto formal wichtiger war und bis dato fehlte, war der Wetterschutz. Sicher, die Kutsche war Vorbild und Hilfsgerät, motorgetrieben als Automobil zu gelten. Aber mit der weiteren Verbreitung des Autos sollte nicht mehr die unsägliche Verkleidung der Passagiere dem Wet-

20. 1916 ist der Detroit Electric wetterfest mit bester Rundumsicht. Trotz E-Motor verrät das Differential an der Hinterachse die Antriebsart. Die flache Fronthaube kündet vom platzsparenden Elektromotor. Die komplett blaue Karosse fördert den optischen Zusammenhalt der dreivolumigen Bauart mit Fronthaube, Aufbau und Heckhaube. (Photo: Wheelsage.)

21. 1922 wußte Lancia bei seinem Lambda, daß Material und Gewicht bei der Stahlverwendung durch Aussparungen in unbelasteten Chassisbereichen reduziert werden kann. Zusätzlich baute man dieses Modell für seine Zeit extrem niedrig. (Photo: Lancia.)

22. 1919 sah der Citroen Type A aus wie Dutzend anderer Marken. Kein Karosseriedetail bis auf die Kühlerform ließ Markentypisches erkennen. Die Scheibenräder beruhigten den Gesamteindruck. Die Passagiere waren leidlich wettergeschützt, eher eine Preisfrage denn eine technische. (Photo: Scoriocars.)

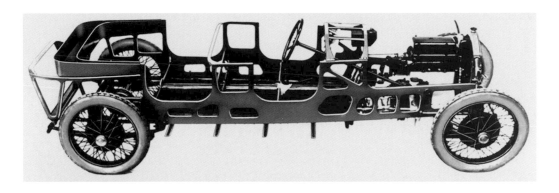

20. The 1916 Detroit Electric is weather tight with a perfect panoramic view. In spite of the electric motor, the differential near the rear axle betrays its type of drive. The flat front hood indicates that there is a space-saving electric motor. The completely blue body promotes the visual cohesion of the three-volume design – a front hood, body and rear hood. (Photo: Wheelsage.)

21. In 1922, when Lancia designed its Lambda, its designers knew that, when steel is used, material and weight can be reduced by means of cutouts in unloaded chassis areas. Moreover, this model was built extremely low for its period. (Photo: Lancia.)

22. In 1919 the Citroën Type A looked like a dozen other makes. Not one bodywork detail except for the shape of the radiator was typical of the Citroën brand. The disc wheels had a steadying effect on the overall appearance. The passengers were protected from the weather reasonably well – more a question of price than one of technology. (Photo: Scoriocars.)

from its modern equivalents, which are laden with switches. The malodorous, carbide fueled headlight was gradually replaced by an electric one, which helped to prevent possible fires in the case of accidents, for after 1908 this became the standard. The car of that era was an expression of affluence, for only the well-to-do could afford the prices, which reflected craftsmanship and heavy material. For instance, a set of pneumatic tires in 1900 cost the equivalent of 10 000 euros. Henry Ford, in the U.S., had taken on the job of changing this. He realized that the more inexpensive a car was, the more cars could be sold. The success of the Model T (Tin Lizzie) – from 1908 to 1927 – was a given. Up to that time it was one of Ford's cost cutting measures that the car would be sold only in black. This color was popular, cheap to produce, fast drying and convenient for the suppliers, for they needed to deliver everything only in black. In the meantime the internal combustion engine had prevailed over other engines. As late as 1900 there were 1500 electric cars, 751 steam-driven cars, but only 15 gasoline-driven cars in the United States. Steam lost out because it was too involved to operate, while the battery was phased out because of its weight and range. On the other hand, the internal combustion-engine technology quickly made up for lost time and has been able to assert its dominance up to the present day – though only up to the present day. For the future probably belongs to the electric motor.

What was formally more important for the car and had hitherto been missing was protection from the weather. Certainly the carriage was a helpful model when it came to developing a motor-driven automobile. But as cars became more popular, they had to have roofs so that passengers would no longer have to dress up in ridiculous motoring clothes in order to brave the weather. One early solution was the folding roof, which, however, could not be used in cars that had a rear entrance (example: the tonneau). Therefore weatherproof leather upholstery also made sense, usually in the form of diamond upholstery, or, more rarely, a fluted design (rounded upholstery with straight lines). The permanently mounted roof, which partially folded back over the driver (in the

23. 1911 muß der Miller Offenhauser »Golden Submarine« (hier ein Nachbau) wie von einem anderen Stern gewirkt haben. Unsere Augen lesen eher eine rasante Zukunftsversion, nur gestört von dem zeitgenössischen planen Kühler. (Photo: Volumeone.)
24. 1911 hatte der Gregoire Torpille aerodynamisch falsch gedacht mit dem runden Heck, wirkte aber höchst fortschrittlich. Fast ein Spaß: Hinter dem runden Wulst verbarg sich das Reserverad. (Photo: Theoldmotor.)
25. 1925 hatte der Hanomag mit Spitznamen »Kommißbrot« unten herum aerodynamisch fast alles richtig gemacht. Der Aufbau dagegen suchte Anleihen am Hausbau. (Photo: K. Meijers.)
26. 1922 soll das Auto erschwinglich werden, auch wenn gespart werden muß und das Ergebnis eher ein Motorrad mit Dach ist: das Mauser-Einspurauto mit winzigen Stützrädern für Start und Stopp. (Photo: Speyer Technikmuseum.)
27. 1931 ist das Auto erschwinglich geworden, so der DKW F1. Auch er nutzt den Kühler zur Wiedererkennung, hier mit einem Knick in der Fläche. (Photo: RM-Auctions.)
28. 1922 war die Bootsform immer noch eine beliebte Karosserievariante. Der Amilcar Type 4c Skiff war von seltener Handwerkskunst des Holzkörpers und der perfekten Kotflügelformen. (Photo: Favcars.)

ter trotzen, Dächer mußten her. Dabei war zunächst das Faltdach ein Weg, das aber bei Karosserien mit Einstieg im Heck (Beispiel: der Tonneau) nicht verwendet werden konnte. Sinnvoll war daher auch die wetterfeste Polsterung mit Leder, meist ausgeführt als Rautenpolsterung, seltener als Pfeifenpolsterung (eine wulstförmige, geradlinige Polsterung). Das fest montierte Dach, teilweise über dem Fahrer wegklappend (Bauweise des Coupé de Ville), teilweise über den Fondfahrgästen wegklappend (Bauweise des Landaulet), wurde wie die Seitenteile zunächst auf einem Holzgeripppe aufgezogen, das teilweise mit Stoff bezogen oder mit Holztafeln bekleidet war. Wetterfest wurde alles mit Lackierung in mehreren Schichten, ein zeitraubendes, teures Verfahren, denn Blech war zunächst nur für Kotflügel und Motorhauben bestimmt. Diese auf Holz basierende Rahmenkonstruktion hatte unverformbare Verbindungen, was sich im Betrieb oft durch Knartzen bemerkbar machte, weil die Karosserie fest mit dem Fahrgestell verbunden und die Straßenqualität mäßig war. Das Weymann-Patent schaffte ab 1922 Abhilfe durch flexiblere, mit Blechstegen verbundene und mit Ölpapier gedämpfte Kontaktflächen des Gerippes. Diese Bauart konnte sich nicht lange halten. Stahl war die Zukunft. Blechkarosserien waren schneller herzustellen, konnten dank neuer Lackrezepturen in Öfen ihre Farben schneller einbrennen und dienten bald ab Mitte der 1920er Jahre für die selbsttragende Karosserie, die der Lancia Lambda 1922 vorbildlich in Szene setzte. Trotz der windfeindlichen, rundum senkrechten Fensterflächen war seine niedrige Bauart selbst für das heutige Auge ein wichtiger Schritt zum modernen Auto.

Das »Kommißbrot« von Hanomag aus dem Jahre 1922 hatte sich ebenfalls noch den senkrecht kantigen Aufbau gegönnt, zeigte aber im Unterbau mit gerundetem Blechbug, zentralem Scheinwerfer-Zyklopenauge, den kaum noch auffälligen, im Grundkörper integrierten Kotflügeln und dem runden Heckabschluß den Weg zur geschlossenen Form auf. Dies trieb der Rumpler Tropfenwagen, ebenfalls von 1922, noch weiter. Kenntnisse der Aerodynamik hatten die Auffassung, wie ein Auto auszusehen habe, grundlegend verändert, was seinen sehr mäßigen Erfolg erklären könnte. Auch schön, aber nicht mehr zeitgemäß gerieten jene Karossen, die sich mit Bootsheck, meist handwerklich treu in Holz, zeigten, wie auch andere mit im Heck speziell gerundeten Formen, die aerodynamisches Profil suchten, was, physikalisch gesehen, verkehrt war. Beispiele hierzu sind der französische Grégoire Torpille von 1911, der amerikanische »Golden Submarine« von Miller-Offenhauser von 1911 oder sein Landsmann Amilcar Type 4c Skiff von 1922.

Neuartige Techniken im Detail bestimmten weiterhin die automobile Form. Die horizontal noch geteilten Frontscheiben waren teilweise hochklappbar, weil Regen die Sicht verschlechterte und erst der elektrische Scheibenwischer (Bosch-Patent 1926, aber schon früher erfunden) die feststehende Frontscheibe ermöglichte, die ab 1923 als Verbundglasscheibe Verletzungen reduzieren konnte. Denn die Kehrseite des Autoerfolgs sah bei uns bereits 1912 noch so aus: 442 Verkehrstote bei einem Autobestand von 70006 Fahrzeugen, das wären 631 Tote auf 100 000 Personen. Wie sehr auch die technische Entwicklung vorankam und damit eine gesteigerte Sicherheit einherging, zeigen diese Zahlen: 1938 230 Tote je 100 000, 1970 (Nachkriegshöchststand) 114 Tote je 100 000 Personen. Knautschzone und Sicherheitszelle kamen erst Ende der 1950er Jahre auf. Tempo war weiterhin das geförderte und beherrschende Thema. Das Auto konnte bereits die 100-km/h-Marke als Standard erfüllen. Da ab ca. 60 km/h der Luftwiderstand am größten von allen Fahrwiderständen ist, blieb es nicht aus, daß die Karosserie beginnen mußte, sich unter dem Wind zu ducken, was auch dem Benzinverbrauch zugute kam. Dieser wurde an Tankstellen kompensiert, zum ersten Mal ab 1922 in Hannover mit verkauften Kanistern und zum ersten Mal ab 1926 in Hamburg mit einer Zapfsäule.

Das Auto, ein teures Transportmittel, forderte kreative Köpfe, die dem Kleinverdiener einen kostengünstigen Weg zur Arbeit und in die Freizeit erleichtern wollten. Zweirädrige Mobile wie ein

Coupé de Ville model), and partially folded back above the backseat passengers (in the Landaulet model), was first mounted, like the side sections, on a wooden framework which was covered with fabric or clad with wooden panels. It was all made weatherproof with several layers of varnish, a time-consuming, expensive process, for metal was initially used only for mudguards and hoods. This wood-based frame construction had non-deformable joints, which creaked when the car was in operation, because the bodywork was permanently attached to the chassis and the roads were in rather poor shape. As of 1922, the Weymann patent remedied the situation by making the contact surfaces of the chassis more flexible, connecting them with metal webs and muffling them with oiled paper. This design was not sustainable for long. The future was steel. Metal car bodies could be manufactured faster, the paint, thanks to recently developed varieties, could be baked on more quickly in ovens, and starting with the mid-1920s they were soon used for the self-supporting body that the 1922 Lancia Lambda featured in such exemplary fashion. In spite of the fact that all its windows are vertical and cause wind resistance, its low-slung design, even to our eyes, is an important step toward the modern car.

The 1922 »Kommissbrot« model by Hanomag still had a vertical, angular body, but its undercarriage, with its rounded steel nose, central Cyclops-eye headlamp, barely noticeable mudguards integrated in the main body and its round rear section, pointed the way toward a self-contained form. The Rumpler Tropfenwagen, also from 1922, went one step beyond this. A knowledge of aerodynamics had fundamentally changed people's idea of what a car should look like, which might explain its very moderate success. Also beautiful, though no longer up-to-date, were limousines that sported a boat tail, usually carefully handcrafted from wood, or others that had specially rounded forms in the rear, seeking to create an aerodynamic profile, which, from the point of view of physics, was wrong. Examples of this are the 1911 French Grégoire Torpille, the 1911 American »Golden Submarine« by Miller and Offenhauser or the 1922 Amilcar Type 4c Skiff, also U.S.-made.

Novel detailing technologies continued to determine the form of the automobile. The windshields, which were still horizontally divided, could be partially folded back, because rain impaired visibility, and it took an electric windscreen wiper (patented by Bosch in 1926, but invented earlier) to make possible the fixed windscreen, which after 1923 was able to reduce injuries, being made of laminated glass. For there was a downside to the car's success here in Germany, as early as 1912: 442 traffic fatalities with a total number of 70,006 cars, i.e., 631 fatalities per 100 000 persons. The following numbers show how far technical development has advanced, bringing with it increased safety: In 1938, there were 230 dead per 100 000, while in 1970 (a postwar record high) there were 114 dead per 100 000 persons. The crumple zone and safety cell were not introduced until the late 1950s. Speed continued to be the dominant topic that was promoted. Cars were already able to reach a standard speed of 100 km/h. Since roughly 60 km/h air resistance is the major rolling resistance, it stood to reason that the car's body had to start to duck under the wind, which also helped reduce fuel consumption. Gasoline was sold at filling stations – the first time, beginning in 1922, in Hannover, where canisters were for sale, and as of 1926 in Hamburg, where they had the first gas pump.

The car, an expensive means of transportation, required creative minds who would make it easier for low-income people to travel cheaply to work and on holidays. Two-wheel vehicles such as the 1922 Mauser Einspurauto (Monotrace), which had small retractable stabilizing wheels, the 1922 Hanomag Kommissbrot, which for the purpose of weight reduction also had a wickerwork shell for its sports cars, the BMW Dixi, a licensed production of the British Austin 7, or the German DKW F1, built as of 1931, tried to fill this niche. Incidentally, the F1 was as modern as the Mini, which was built three decades later: it had a transverse front-mounted engine with front

23. In 1911 the Miller Offenhauser »Golden Submarine« (pictured here is a replica) must have looked like something from another planet. We, however, see it as a dynamic, futuristic version, marred only by the flat radiator of its period. (Photo: Volumeone.)
24. The 1911 Grégoire Torpille was aerodynamically wrong because of its round rear, but looked extremely progressive. What was almost funny was that the round bulge concealed the spare wheel. (Photo: Theoldmotor.)
25. From an aerodynamic point of view, the lower part of the 1925 Hanomag, nicknamed »Kommissbrot« (after the square loaf of army bread), had almost everything right. But the body reminded you of a house. (Photo: K. Meijers.)
26. In 1922 cars were supposed to become affordable, even if you had to skimp and the result was more like a motorbike with a roof: the Mauser Monotrace with tiny support wheels for starts and stops. (Photo: Technikmuseum Speyer.)
27. Cars have become affordable, e.g., the 1931 DKW F1. It, too, uses the radiator for brand recognition, here with a crease in its surface. (Photo: RM-Auctions.)
28. In 1922 the boat shape was still a popular body version. The craftsmanship of the wooden body and of the perfect forms of the fenders of the Amilcar Type 4c Skiff was outstanding. (Photo: Favcars.)

Mauser-Einspurauto von 1922 mit einklappbaren Stützrädchen, der Hanomag Kommißbrot von 1922, der zwecks Gewichtseinsparung auch für sportliche Einsätze eine Hülle aus Korbgeflecht hatte, der BMW Dixi als Lizenzproduktion des englischen Austin 7 oder der ab 1931 gebaute deutsche DKW F1 waren bemüht, diese Lücke zu füllen. Übrigens war der F1 so modern wie der drei Jahrzehnte später gebaute Mini: mit querliegendem Frontmotor mit Frontantrieb (erstes deutsches Auto mit diesem Antrieb), allerdings immer noch mit Kunstleder bezogener Holzkarosse. Die Mittelklasse und alle Modelle darüber hatten sich inzwischen auf die selbsttragende Stahlkarosse in Fließbandproduktion geeinigt. Die USA gönnten sich bei General Motors 1927 das erste Designstudio, damals noch »Art and Colour Section« genannt. Manche verspotteten es als Geburtshelfer der »planned obsolescence«, der planmäßigen Frühveralterung gängiger Modelle zugunsten neuer. Viele Entwürfe wurden jedoch nicht geplanter Frühveralterung geschuldet, sondern vom Wind beflügelt. Modelle wie ein Maybach Zeppelin von 1932 (als »heimlicher« Vorläufer der späteren Pontonform), ein Aerotatra V 570 von 1933, der Chrysler Airflow von 1934, die Modelle Opel Kadett, Cord 810 und die Toyota A-Serie von 1936 sind Beispiele, wie sich das Bild des Automobils über dreißig Jahre veränderte. Aus Einzelelementen, die das additive Prinzip aller Anbauteile optisch demonstrierten, bildete sich mit der Entwicklung aller Details die geschlossene Karosserieform. Fondsitze rückten dank tiefer liegender Fahrgestelle vor die Hinterachse. Die Motoren wurden wegen kleinerer Hubräume niedriger gebaut. Die elektrische Beleuchtung konnte kleiner gemacht werden als ihre azetylengefütterten Vorläufer. In der Summe bot die Automobilform langsam dem Wind weniger Widerstand. Die einen integrierten in die Grundform Details wie frei stehende Scheinwerfer oder die gerundete Kühlerattrappe mit dem dahinter liegenden »Echt«-Kühler. Andere, wie die Hersteller der Tatras, studierten die Flugzeug-Aerodynamik. Und schon in den 1920er Jahren zeigten Spezialkarossen einiger Hersteller die aerodynamischen Erkenntnisse von Paul Jaray, später von Wunibald Kamm und von Reinhard von Koenig-Fachsenfeld und anderen Erfindern, so des Graphikers Ernst Neumann-Neander mit neuen aerodynamischen Formen für seine »Fahrmaschinen«. Le Corbusier hatte 1936 mit einem Entwurf das Prinzip des preisgünstigen »voiture minimum« gezeigt (erste Konzepte seit 1928), dem gedanklichem Vorläufer des Käfers. Der amerikanische Architekt Richard Buckminster Fuller glaubte 1933 an den

29. 1936 machte der Cord 810 Westchester Sedan Furore mit Klappscheinwerfer, Verzicht auf sichtbaren Wabenkühler zugunsten feiner Chromstreifenapplikationen und niedriger Gesamtbauhöhe. (Photo: Wheelsage.)
30. 1936 konnte der Opel Kadett eine selbsttragende Karosserie zur Gewichtseinsparung als Vorteil nennen. In die Grundform integrierte Scheinwerfer und eine gerundete Kühlermaske sah man auch bei großen Opelmodellen und bei Renault. Rußland profitierte von Reparationszahlungen nach dem 2. Weltkrieg: Aus dem Kadett wurde ein formalidentischer Moskwitsch. (Photo: Opel.)
31. 1933 sah der Aero-Tatra V 570 noch etwas hilflos aus als ein windschlüpfiges Beispiel. Seine Nachfolger holten schnell auf und wurden noch glatter mit gewölbten Frontscheiben, was den Luftwiderstand weiter reduzierte. (Photo: Pinterest.)
32. 1934 war sein Name Gestaltungsprogramm: Chrysler Airflow Eight Sedan. Runde Front, glattflächige Karosse, verkleidete hintere Radausschnitte, der Weg zur aerodynamisch korrekten Form wird frei. (Photo: RM-Auctions.)

20

29. The 1936 Cord 810 Westchester Sedan created quite a stir: while it had pop-up headlights and no visible honeycomb radiator, it featured delicate chrome strips and an overall lowslung body. (Photo: Wheelsage.)

30. In 1936 one of the advantages of the Opel Kadett was a unitized body to reduce weight. Headlights integrated in the basic form and a rounded radiator grille were also seen in large Opel models and in Renault vehicles. Russia profited from reparation payments after World War II: The Kadett turned into the formally identical Moskvich. (Photo: Opel.)

31. For a streamlined design, the 1933 Aero-Tatra V570 still looked somewhat helpless. Its successors caught up quickly and became even sleeker, with curved windshields, which further reduced aerodynamic resistance. (Photo: Pinterest.)

32. It was the latest in design: the 1934 Chrysler Airflow Eight Sedan. A rounded front, slipstream body and hidden rear wheel cutouts paved the way to an aerodynamically correct form. (Photo: RM-Auctions.)

drive (the first German car with this drive), though the coachwork was still made of wood covered with synthetic leather. The mid-size class of cars and all models above it had in the meantime changed over to self-supporting steel bodies manufactured on the assembly line. In 1927, the U.S. now had the first design studio at General Motors, which was still called the »Art and Color Section«. Some people made fun of it because it gave rise to the »planned obsolescence« of current models in favor of new ones. But many designs were not so much influenced by planned obsolescence, but inspired by the wind. Designs such as a 1932 Maybach Zeppelin (a »secret« precursor of the later pontoon form), a 1933 Aerotatra V 570, the 1934 Chrysler Airflow, the models Opel Kadett, Cord 810 and the Toyota A series of 1936 are examples of the way the appearance of the car changed over thirty years. As all the details were developed, out of individual elements that visually demonstrated the additive principle of all add-on components, was formed the integrated body design. Thanks to a lower-slung chassis, backseats moved in front of the rear axle. Because of smaller cylinder capacities, engines were built lower. Electric lights could be made smaller than their acetylene-fueled precursors. On the whole, the form of the car gradually came to offer less resistance to the wind. Some designers integrated into the basic form details such as free-standing headlights or a rounded mock radiator with the »genuine« radiator behind it. Others, like the manufacturers of the Tatras, studied the aerodynamics of airplanes. And as early as in the 1920s, the special car bodies of some manufacturers showed the results of the aerodynamic research of Paul Jaray, later of Wunibald Kamm and of Reinhard von Koenig-Fachsenfeld and other inventors – for instance, the graphic artist Ernst Neumann-Neander, with new aerodynamic forms for his »Fahrmaschinen« (driving machines). In a 1936 sketch, Le Corbusier had shown the principle of the inexpensive »voiture minimum« (the first concepts date back to 1928), the theoretical precursor of the Beetle. The American architect Richard Buckminster Fuller, in 1933, believed in the wingless aircraft body on three wheels: a prototype blueprint for recipes for failure. A car must not be focused on the few advantages of its concept. On the other hand, how groundbreaking, in 1913, was Count Ricotti's car: Built by Castagna on an Alfa Romeo chassis, a mixture of car body and Zeppelin, it featured a huge panoramic windshield subdivided by

flügellosen Flugzeugkörper auf drei Rädern: ein prototypischer Entwurf für Rezepte des Scheiterns. Ein Auto darf sich nicht auf wenige Vorteile seines Konzepts konzentrieren. Wie wegweisend war es dagegen bereits 1913, als Graf Ricotti auf einem Alfa-Romeo-Fahrgestell ein Auto zeigte, das, von Castagna gebaut, als Mischung aus Karosserie und Zeppelin auffuhr: riesige, mit Stegen unterteilte Panorama-Frontscheibe mit ausgerundeten Ecken, seitliche Bullaugen, ausgerundete Türen, was verarbeitungstechnisch klüger war als rechteckige Ausschnitte für Fenster und Tür. Größter Nachteil: Der Bug hatte eine große, aerodynamisch ungünstige Öffnung für den Kühler, am Fahrerplatz wurde es ziemlich heiß, und der Raum zum Motor war kaum abgeschirmt. Heilung brachte die »Sommerversion«: alles wie gehabt samt großer Frontscheibe, aber ohne Dach, optisch alles andere als geplante Obsoleszenz. Auch der Architekt Walter Gropius versuchte sich am Autodesign. 1932 verpaßte er den Adler-Modellen Super 6 und Super 8 klare, aber konservative Karosserien auf Rädern mit flächig gestuften, verchromten Radkappen und einem Kühler im Stil eines Rolls-Royce, aber oben abschließend anstatt tempelartig mit weichem Bogen.

1938 wurde das Fundament für den KdF-Wagen, den Volkswagen, gelegt. Die Nationalsozialisten wollten ein Volksauto schaffen, das deutsche Bürger mit ihrem Ansparsystem kaufen sollten. Ähnlich wie schon ein Mercedes W17 von 1931 hatte der Käfer, so sein Kosename nach dem Krieg, Heckmotor und damit die Chance für einen kühlerlosen gerundeten Bug wie der W17. Ohne sichtbaren Kühler, aber mit Frontmotor hatte der Saab 1946 die Käferform noch konsequenter windschlüpfig gestaltet und diese Form prinzipiell lange, aber nicht so lange wie der VW Käfer, beibehalten. Im Krieg wurde die zivile PKW-Produktion fast weltweit eingestellt. Aber ausgerechnet in dieser Zeit hatte der US-Jeep von 1941 für Friedenszeiten das Vorbild geliefert für aktuelle, hochmoderne SUV-Modelle (SUV = Sport Utility Vehicle). Selbst seine zunächst fünf und dann sieben senkrechten Luftschlitze im Bug sind heute, modisch gestaucht, immer noch die Visitenkarte der Marke, ebenso wie die Niere bei BMW und die Kühlerattrappe bei Mercedes. Den Neuanfang bei uns bildeten Autos mit Karossen aus der Vorkriegszeit: Opel Olympia, Ford Taunus (Buckel-Ford), Mercedes 170 V. 1948 war es in Frankreich die »Ente«, der Citroën 2CV, bereits 1938 mit mehreren hundert Prototypen getestet – ein Auto, bei dem weniger mehr war. Borgward aber war 1949 mit dem Hansa 1500 das erste deutsche Auto mit Pontonkarosserie. In den USA versuchte sich 1948 die kleine Marke Tucker mit fortschrittlicher Technik: Servolenkung, Sicherheitsgurte, Kurvenlicht in Bugmitte, Sechs-Zylinder-Boxermotor und windschlüpfige Karosserie. Die drei Großen (GM, Ford und Chrysler) waren aber auf dem Weg zu Chrom, viel Blech und wachsenden Heckflossen. Kein Großer, aber etwas mutiger war Nash mit seinen extrem glattflächigen, voll überdeckten Rädern und (fast) vollendeter Pontonform, aber ohne Zukunft. Die begann für viele erst jetzt so richtig.

33. 1938 war der Käfer gar nicht mal spektakulär. Immerhin hatte sein Konstrukteur Ferdinand Porsche schon vorher formal ähnlich gestaltete PKWs entwickelt, auch mit Heckmotor und glatt gerundetem Blechbug. (Photo: Volkswagen AG.)

34. 1931 entwickelte Ferdinand Porsche den Mercedes W17. Schon hier kündigt sich der Käfer an, denn der Blechbug ohne Lüftungsschlitze verriet den Heckmotor im fließend auslaufenden Heck. (Photo: Daimler.)

35. 1941 wird ein Klassiker geboren, allerdings für den Kriegseinsatz und ein Vobild für fast alle Sport Utility Vehicles, kurz SUVs: der Jeep von Willys mit Vierradantrieb, das wohl unverwüstlichste Fahrzeug aller Zeiten. (Photo: Willys.)

36. 1948 war der Tucker Torpedo formal unausgeglichen, machte es wett mit technischer Raffinesse: windschlüpfige Form (Cw-Wert 0,27), Servolenkung, Sicherheitsgurte, Kurvenlicht im Zentralscheinwerfer, 6-Zylinder Boxer-Heckmotor. Gerade 51 Autos kamen auf amerikanische Highways. (Photo: P. Litwinski.)

37. 1949 hatte die kleine US.Marke Nash den Mut, alle vier Räder fast komplett zu verstecken, was der Aerodynamik gut tat. Gesicht und Aufbau waren weniger mutig, der Erfolg war mäßig, das Überleben der Marke hielt nicht lange. (Photo: Autopics.)

33. The 1938 Beetle was not spectacular. Still, its designer, Ferdinand Porsche, had developed formally similar passenger cars prior to this that also had a rear-mounted engine and smoothly rounded metal nose. (Photo: Volkswagen AG.)

34. Ferdinand Porsche developed the Mercedes W17 in 1931. It already had hints of the Beetle about it, for the metal front end without ventilation slots revealed that the rear-mounted engine was concealed in the sleek tail end. Like the Mercedes W17, its successors 130H, 150H and 170H had only moderate driving stability because the rear engine was installed all the way in the back. (Photo: Daimler.)

35. In 1941 a classic is born, though only for military use, becoming a model for almost all sport utility vehicles, or SUVs: the Willys Jeep with four-wheel drive – probably the most indestructible vehicle of all time. (Photo: Willys.)

36. The 1948 Tucker Torpedo was formally uneven, but made up for it with technical sophistication: a streamlined form (drag coefficient 0.27), power steering, seat belts, cornering light in the central headlight, 6-cylinder Boxer rear engine. A mere 51 Torpedos hit the American highways. (Photo: P. Litwinski.)

37. The little American brand Nash was bold enough in 1949 to conceal all four wheels almost completely, which improved its aerodynamics. Its overall look and body were not as bold, its success was modest, and the brand did not survive long. (Photo: Autopics.)

crosspieces, with rounded corners, portholes on the sides, and rounded doors, which was smarter in terms of production technology than having rectangular window and door openings. Its greatest disadvantage: The nose had a large, aerodynamically unfavorable opening for the radiator, it got quite hot in the driver's seat, and the space for the engine was barely shielded. The »summer version« was a big improvement: everything as before including the big windscreen, but without a roof – anything but planned obsolescence even today, from a visual point of view. The architect Walter Gropius also tried his hand at car design. In 1932 he gave the Adler models Super 6 and Super 8 clear, but conservative bodies on wheels with stepped, chrome-plated hubcaps and a radiator in the style of a Rolls-Royce, but with a soft curve instead of temple-like at the top.

In 1938 the foundation was laid for the KdF [Kraft durch Freude, or strength through joy] car, the Volkswagen. The National Socialists wanted to create a people's car that would be bought by German citizens through a special savings plan. Like a 1931 Mercedes W17, the Beetle, as it was nicknamed after the war, had a rear-mounted engine and could thus – like the W17 – have a rounded nose without a radiator. Without a visible radiator, but with a front-mounted engine, the 1946 Saab had developed an even more streamlined beetle form and basically kept this shape for a long time, but not as long as the VW Beetle. During the war the manufacture of civilian passenger cars was stopped virtually everywhere in the world. But precisely during this period the 1941 American Jeep provided a model for current, ultramodern postwar SUVs (sport utility vehicles). Even its vertical front ventilation slots – initially five, then seven – fashionably swaged, are still the hallmark of the Jeep today, as is the kidney grille of the BMW and the mock radiator of the Mercedes. In Germany, cars with prewar bodies provided a fresh start: Opel Olympia, Ford Taunus (or »Buckel-Ford«, »Hunchback« Ford), Mercedes 170 V. In 1948 French manufacturers introduced the »Duck«, the Citroën 2CV, which had already been tested in 1938 along with several hundreds of prototypes – a car in which less meant more. But Borgward, with its 1949 Hansa 1500, was the first German car with a pontoon body. In the U.S., the small Tucker company in 1948 featured advanced technology: power steering, seatbelts, a front center cornering light, a six-cylinder boxer engine and a streamlined body. But the big three (GM, Ford and Chrysler) were on their way to chrome, lots of metal and ever-increasing tailfins. While it wasn't one of the big three, the Nash with its extremely sleek, fully skirted wheels and an (almost) perfect pontoon form was somewhat bolder, though it had no future. Meanwhile that future, for many, didn't really begin until that point.

3. Von 1950 bis heute. Die Form bestimmt die Technik

Der Zweite Weltkrieg war vorüber, vieles aufgeräumt, die Fließbänder bewegten sich wieder. Selbst individuelle Karossiers in Europa oder den USA fanden ihre Käufer. Verwunderlich, daß Firmen wie Figoni & Falaschi, Delahaye, Pininfarina, Hooper und viele andere in Europa die teuren Automobile absetzen konnten. Bereits 1949 zeigte Borgward mit dem Hansa 1500 die Pontonkarosse in Deutschland, während heimische Konkurrenten in Vorkriegsformen verharrten, wie der Buckel-Ford, ein Opel Olympia mit Ersatzrad auf dem Heck, ein Mercedes mit frei stehenden Scheinwerfern und Trittbrettern. Frankreich war schon weiter bei seinen Vorkriegsmodellen. Ein Citroën 7CV (der »Gangsterwagen«) konnte seit 1934 seine Fahrqualität dank breiter Spur und langem Radstand bis 1955 retten. Peugeot hatte mit seinem Vorkriegs-202 das aerodynamische Gesicht mit hinter Stegen versteckten Scheinwerfern geprägt. Italien, von Haus aus die Heimat talentierter, handwerklich geschickter Karossiers, wagte mit einem Lancia Aprilia oder einem Fiat 1100 den Nachkriegsstart. Der Fiat 1400 war 1950 in klarer Pontonform in der Neuzeit angekommen. Lancia zögerte und versteckte das Pontonprinzip unter eleganter, geneigter Kühlerfront und in angedeuteten Kotflügeln mit integrierten Scheinwerfern.

In den USA war man mutiger. Hatten erste Concept-Cars, Blech gewordene Ideen auf Rädern, wie der 1938 konservativ gezeichnete Buick Y-Job mit angedeuteten Kotflügeln oder der modernere Chrysler Thunderbolt 1941 im Pontonkleid die neuen Formen vorgezeichnet, die der Tucker 1948 glücklos umsetzte, so zeigte Studebaker noch mehr Mut. Raymond Loewy, bekannter US-Designer mit der Referenz der weltbekannten Lucky-Strike-Zigarettenschachtel mit dem roten Punkt und der Coca-Cola-Flasche mit Taille und Hüfte, gab mit dem Studebaker Commander Starlite Coupé 1949 vor, was modern sei. Speziell das in die Seiten gezogene Heckfenster – wir kennen es als Frontfenster vom Alfa Castagna 1913 und ab 1953 sogar beim DKW – machte aus den Aufbauten mit knappen Glasflächen ein lichtdurchflutetes Ambiente, aber mit konservativ gestaltetem Bug. 1950 war das Flugzeugmotiv im Autodesign angekommen. Loewy verpaßte dem Starlite Coupé von 1951 eine kreisrunde große Nase mit verchromter Spitze. Die Phantasie dachte sich Propeller oder den gerade aufkommenden Düsenantrieb der Kampfjets dazu. Das riesige Heckfenster blieb unverändert. Trotz hinterer Kotflügelmotive war dieses Auto Beispiel für die neue Zeit, denn GM, Ford und Chrysler als die großen Serienhersteller waren weniger mutig. Als hätten sie den Ausspruch von Loewy beherzigt, der das MAYA-Prinzip für Design postulierte: »Most advanced yet acceptable« – so modern, um gerade noch akzeptiert zu werden. Eine Lehre, die bereits gut war für die 1930er Jahre, als aerodynamisch optimierte Karosserien auf den Markt kamen. Beispiele wie ein Chrysler Airflow oder der Adler 2,5 Liter Autobahn waren den meisten Augen zu futuristisch, um sich den Erfolg zu sichern.

Bei uns kamen 1950 der Porsche 356a als Stuttgarter Produktion (den Vorläufer in kleiner Auflage lieferte Österreich seit 1948) und als reines Nutzfahrzeug der VW Transporter. Beide ein Porsche-Entwurf, beide mit luftgekühltem Heckmotor, beide mit Details preiswerter Auslegung von Antrieb und Fahrgestell. Der 356a war glattflächig, weich und rund geformt, was den 40 PS eine Höchstgeschwindigkeit von 135 km/h schenkte. Dem Transporter langten maximale 85 km/h. Prägend war seine Front mit dem großen Kreis des VW-Logos, eingerahmt in zwei aus den Seiten laufenden, ins Blech geprägten Linien, manchmal farblich von der Grundform abgesetzt. Praktisch für die VW-Werkstätte: Der T1 hatte mit 240 cm denselben Radstand wie der Käfer. Nachteil als Transporter: Der Heckmotor war wegen der hohen Ladekante von hinten schwer zu beladen. Das machten die Konkurrenten Tempo Matador (anfänglich mit VW-Motor in Mittelmotorlage) und der DKW Schnellaster und Jahre später der Ford Transit (beide mit Frontmotor) besser, aber ohne ähnlichen Erfolg wie beim T1. Bei neuen Antriebsarten dachte man 1950 an Rover in England. Nicht die Serienautos interessierten, sie waren wie viele Marken der Insel konservativ gestaltet, sondern Rovers Versuch, eine Gasturbine zu verwenden – eine Idee, die Chrysler 13 Jahre später ebenso erfolglos aufgriff. Der Rover, formal ein Cabrio in englischer Zurückhaltung als Pontonkarosse, hatte bei seiner Blechhaut nicht ahnen lassen, daß im Inneren Kolbenloses werkelte. Der Chrysler zeigte 1963 mehr Mut mit seinen Turbinemotiven an Bug und Heck.

Am unteren Ende der fahrbaren Möglichkeiten bot Borgward 1951 den Lloyd 300. Die Pontonform war ideal für eine Holzkonstruktion aus mit Kunstleder bespanntem Sperrholz, aufgesetzt auf einen zweitaktik angetriebenen Zentralrohrrahmen. Der Spruch »Wer den Tod nicht scheut, fährt Lloyd« oder sein Spitzname »Leukoplastbomber« konnten seinen Erfolg nicht beeinträchtigen, er wurde stufenweise zur reinen Blechkarosse mit 600 ccm und machte am Ende mit 320 cm Länge und 15-Zoll-Rädern ein vollwertiges Auto aus, das erst mit der Borgward-Pleite 1961

38. 1934 bis 1955 kann sich der Citroen 11CV dank bester Fahreigenschaften halten. (Photo: Citroen.)
39. 1938 macht sich der Peugeot 202 windschnittig. (Photo: B. Werner.)
40. 1949 der Cadillac Series 61 mit Heckflossenwarzen. (Photo: Momentcar.)
41. 1951 Der Studebaker Commander Starliner versucht sich mit Propellernase (Photo: Cargurus.)
42. 1949 hatte Ferdinand Porsche nach dem Krieg zunächst in Gmünd (Österreich) die ersten Autos gebaut. Das Modell 356-2 hatte noch die geteilte Frontscheibe, hilflose Chromstreifen zur Vermeidung optischer Leere im Bug, aber die klassisch glatte Grundform, die 1965 beim 911 die Perfektion erreichte und bis heute formaler Maßstab (fast) aller Porsche Sportwagen ist und bleiben wird. (Photo: Wordpress.)

38. The Citroën 11CV keeps going from 1934 until
1955 thanks to excellent handling characteristics.
(Photo: Citroën.)
39. The 1938 Peugeot 202 goes streamlined. (Photo: B. Werner.)
40. The 1949 Cadillac Series 61 with its wart-like
tail fins. (Photo: Momentcar.)
41. The 1951 Studebaker Commander Starliner
goes for a propeller nose (Photo: Cargurus.)
42. After the war, in 1949, Ferdinand Porsche initially built his first cars in Gmünd (Austria). The 356-2
model still had the divided front windshield, awkward chrome strips to avoid a visual vacuum in the
front section, but did have the classically smooth
basic form, which reached perfection in 1965 with
the 911 and to this day continues to be the formal
standard of (almost) all Porsche sports cars. (Photo: Wordpress.)

3. From 1950 until today. The form determines the technology

World War II was over, much had been put to rights, the assembly lines were moving again. Even some of the coachbuilders in Europe or the U.S. found their buyers. It was surprising that firms like Figoni & Falaschi, Delahaye, Pininfarina, Hooper and many others were able to sell their expensive cars in Europe. As early as 1949 Borgward with its Hansa 1500 was showing the pontoon body style in Germany, while competitors at home persisted in producing prewar styles, like the »Buckel-Ford«, an Opel Olympia with a spare tire on the back, a Mercedes with free-standing headlights and running boards. France had made a lot more progress with its prewar models. Since 1934 a car like the Citroën 7CV (the »gangster car«) had been able, thanks to its wide track and long wheelbase, to maintain its ride quality all the way to 1955. Peugeot with its prewar 202 had influenced the aerodynamic appearance of cars with headlights hidden behind bars. Italy, the original home of talented, skillful coachbuilders, ventured to launch a Lancia Aprilia or a Fiat 1100 in the postwar period. The Fiat 1400 had arrived in the modern era in 1950 in a distinct pontoon form. Lancia hesitantly concealed the pontoon principle under its elegant, slanted radiator front and hinted-at mudguards with integrated headlights.

In the U.S., designers had more courage. While the first concept cars, ideas turned metal on wheels – like the conservatively designed 1938 Buick Y-Job with minimal fenders or the more modern 1941 Chrysler Thunderbolt with pontoon styling – charted the new forms that the Tucker unsuccessfully implemented in 1948, Studebaker was even more courageous. In 1949, with the Studebaker Commander Starlite coupé, Raymond Loewy, the well-known American designer to whom we owe the world-famous Lucky Strike cigarette box with the red dot, and the Coca-Cola bottle with a waist and hips, defined what was modern. Particularly the wrap-around rear window – we know it as the front window of the 1913 Alfa Castagna and even as a feature of the DKW after 1953 – flooded with light car bodies that had hitherto had only minimal windows, while the design of the front was still conservative. By 1960 the aircraft motif had caught on in car design. Loewy gave the Starlite coupé a large bullet nose with a chrome-plated tip. People who were imaginative mentally added propellers or the newly developed jet propulsion of fighter jets. The huge rear window remained unchanged. In spite of the style of its rear fenders this car was exemplary for the new era, for the large series manufacturers – GM, Ford and Chrysler – were less courageous. It was as if they had taken to heart Loewy's MAYA principle for design: »Most Advanced Yet Acceptable« – just modern enough to still be accepted. It was a doctrine that was already good for the 1930s, when aerodynamically optimized car bodies came on the market. For most eyes such models as the Chrysler Airflow or the Adler 2.5-liter (a.k.a. Autobahn Adler) were too futuristic to be successful.

In 1950 Germany manufactured the Porsche 356a, built in Stuttgart (its precursor had been produced in Austria since 1948 in limited edition), and the VW Transporter, a purely commercial

endete. Der Lloyd war gut genug, als LT 500 seit 1953 und als LT 600 seit 1955 das Prinzip des Vans weit vor den 1980er Jahren eingeführt zu haben, geformt aus Motorhaube und busähnlichem Aufbau, nach dem Two-Box-Prinzip. Am oberen Ende automobiler Angebote zeigte der Opel Kapitän 1951 den Wandel der Vorkriegsform in moderner Umsetzung. Noch herrschten langgestreckte Kotflügel in Ablehnung der Pontonform vor, der Bug aber kaschierte mit kräftig geformten horizontalen Chromstegen die einst hohe Kühlerfront, unterstützt von in die Kotflügel eingelassenen Scheinwerfern. Ganz oben aber setzte der Mercedes 300 1951 neue Maßstäbe, nach dem Motto: Wir haben es geschafft. Die Karosserie hatte den klassischen Kühler, lesbare, aber wenig ausgeformte Kotflügelkonturen vorne und hinten und ein sehr rundes Heck als Klassiker des konservativen, aber gefälligen Automobils, meist in Schwarz, selten in Beige und noch seltener in Weinrot. Die späteren Varianten als Coupé oder als Cabrio waren seltene Beispiele einer gelungenen, ja gekonnten Ableitung einer Limousinen-Grundform. Logisch, daß der damalige Bundeskanzler diesen Wagen wählte, was dem 300er den Namen »Adenauer-Benz« einbrachte. Dagegen anzukämpfen, fiel dem BMW 501 ab 1952 sehr schwer. Eine klassische, hohe Niere, ausschwingende, ausgeprägte Front- und Heckkotflügel mit einem Rundheck, das formal betonter war als beim 300er, gute Fahreigenschaften und sänftengleicher Transport halfen nicht, der 501 blieb der ewig Zweite in der Wirtschaftswunderklasse.

Bei der Mittelklasse der Wirtschaftswunderzeit lockte der Ford 12m 1952 mit einer extrem klar gezeichneten Pontonform. Die Flanken gönnten sich eine lange, blechgeprägte Gerade mit Versatz über den Vorderrädern, dazu angedeutete Heckflossen, weil die Oberkanten des Grundkörpers gerade geführt wurden. Klein, aber mit großer Wirkung war die Nase über zweigeteiltem, rechteckigen Grill: Die Weltkugel prangte mittendrin, ein Studebakermotiv en miniature als unübertriebener USA-Verweis. Darauf verzichtete der große Borgward Hansa 2400. Sein Mut lag in dem in der Luxusklasse unüblichen Fließheck – eine angenäherte Form des Aerodynamikers Kamm, der damit in den 1930er Jahren den Kompromiss zwischen sinnvoller Raumausnutzung und optimaler Aerodynamik suchte. Eine panoramaähnliche Frontscheibe mit positiv geneigter A-Säule, drei Seitenfenster, ausgeprägte Betonung der Radausschnitte, verbunden mit formidentischem Schweller dazwischen, ein Motiv der späteren Isabella, und der große Rhombus im Bug als weithin erkennbares Markenzeichen brachten keinen Erfolg. Die nachgeschobene Stufenversion blieb ebenso hilflos hinter dem 300er oder dem 501 zurück.

1953 punktete die GM-Tochter Opel mit dem neuen Rekord. Die Ponton-Grundform wurde belebt durch die backenförmige Wölbung über den Hinterrädern. Auffallend war aber der Bug mit der breiten Öffnung und differenzierenden Stegen auf dem Basisblech, ähnlich einem Haifischmaul. Richtige Zähne zeigte der aus dem GM-Konzern stammende Chevrolet Corvette. Zwei Dinge machten ihn weltbekannt: die Kunststoffkarosse und die europäisch anmutende Eleganz eines Roadsters ohne amerikanische Chromfülle. Auf die verzichtete 1953 der brave Mercedes 180. Seine in die Pontonflanken geprägten Linien früherer Kotflügelprofile waren Jahrzehnte später bei der E-Klasse wieder auferstanden als mißverstandene Maxime des Chefdesigners Bruno Sacco in den 1980er Jahren. Er sprach von vertikaler Affinität und horizontaler Homogenität, kurz: Ein Benz muß als Benz erkennbar bleiben. Das kann dem SL 300 von 1954 attestiert werden: dem Rennwagen ohne Kühler, aber mit typischem, Stern tragendem Bug und dem ewig gültigen Bild der Flügeltüren, unabhängig von vielen anderen, die diese Türtechnik nutzten. Zu Recht haben weltbekannte Designer den SL 300 zum Auto des Jahrhunderts gekürt. Ein Jahr zuvor war

43. 1952 kann der Ford Taunus 12m die Pontonkarosserie gefälliger zeigen als drei Jahre zuvor der Borgward Hansa 1500. Vorteil der klaren Flächen- und Volumengliederung: üppiger Raum für Passagiere und Gepäck. Opel zog erst 1953 nach mit einer pausbäckigen Pontonvariante (Photo: Ford.)

44. 1955 nutzte der Lloyd LT 600 die Technik seiner Limousinenvarianten. Frontmotor mit Frontantrieb, dahinter nur noch Platz und fertig war der erste deutsche Van mit Platz für sechs Personen. Sein Gesichtsausdruck ähnelte dem Tempo Matador von 1950, beide einem Boxerhund nicht unähnlich. (Photo: Gwafton.)

45. 1952 hatte Borgward den Mut, ein Schrägheckauto in der Oberklasse zu positionieren. Der Hansa 2400 war ohne Erfolg, seine Stufenheckversion danach konnte auch nicht mehr helfen. (Photo: L. Spurzem.)

46. 1953 begann der Siegeszug der Baureihe Mercedes W 120 mit seinen Ablegern. Gerade als Taxi wurde er Marktführer. So oder privat genutzt, nie suchte er den Anflug von Extravaganz. Deutsche Bürgerlichkeit in Blech gestanzt. (Photo: Daimler.)

47. 1954 hauchte der VW Karmann Ghia italienische Eleganz in deutsche Länder. Wer die Handschrift des Karossiers Ghia kennt, findet die (fast) zeitlose Form (fast) ewig gültig. (Photo: Volkswagen AG.)

43. The 1952 Ford Taunus 12m shows off the pontoon body more attractively than did the Borgward Hansa 1500 three years earlier. The advantage of a clear-cut arrangement of surfaces and volumes: plenty of room for passengers and luggage. Opel did not produce its chubby-faced pontoon version until 1953. (Photo: Ford.)

44. In 1955, the Lloyd LT600 used the technology of its limousine versions. A front engine with a front-wheel drive, with plenty of space behind it, and there you have it – the first German van with room for six person. Its facial expression resembled that of the 1950 Tempo Matador, both not unlike a boxer dog. (Photo: Gwafton.)

45. In 1952 Borgward had the courage to position a hatchback in the top class. The Hansa 2400 was not successful, and its later notchback version came too late to be any help. (Photo: L. Spurzem.)

46. 1953 marked the beginning of the triumphal march of the Mercedes W120 series with its spin-offs. It was as a taxi that it became a market leader, but even when used for private purposes it never tried to be extravagant. German middle-class culture stamped in metal. (Photo: Daimler.)

47. The 1954 VW Karmann Ghia brought a touch of Italian elegance to Germany. Those who are familiar with the signature of the coachbuilder Ghia will find that its (almost) timeless form is (almost) eternally relevant. (Photo: Volkswagen AG.)

vehicle. Both were Porsche designs, both had an air-cooled rear-mounted engine, both had inexpensively designed motor and chassis details. The 356a was smooth-surfaced, with soft, round contours, which gave its 40 hp a maximum speed of 135 km/h. The Transporter reached a maximum speed of only 85 km/h. Its characteristic feature was the front with the big circle of the VW logo, framed between two lines that ran from the sides, stamped into the metal, sometimes in a color that contrasted with that of the rest of the body. Practical for the VW workshop: The T1, with 240 cm, had the same wheelbase as the Beetle. Its disadvantage as a van: The rear-mounted engine was difficult to load from the back because of the high loading sill. The competitors of T1, Tempo Matador (initially with a mid-mounted VW engine) and the DKW Schnellaster, and years later the Ford Transit (both with a front-mounted engine) did things better, but without the same success as T1. When people had new types of drive in mind in 1950, they thought of Rover in England. It was not the series-manufactured cars people were interested in, for like many British brands they were conservatively designed, but Rover's experiment with a gas-turbine engine – an idea that Chrysler was to pick up 13 years later with equal lack of success. With typical British restraint, the sheet-metal body of the Rover, formally a cabriolet with a pontoon body, gave no indication that its interior was piston-free. The 1963 Chrysler showed more courage with its turbine-like elements in front and back.

Borgward, in 1951, offered a low-end option – the Lloyd 300. The pontoon form was ideal for a wooden body – plywood with an outer skin of synthetic leather, placed on a central tube frame with a two-stroke drive. The saying »Not afraid to die? Drive a Lloyd!« or its soubriquet »Bandaid Bomber« (Leukoplastbomber) could not adversely affect its success. Eventually its body came to be made entirely of steel with 600 ccm and, with a length of 320 cm and 15-inch wheels, it finally became a full-fledged car that did not go out of production until 1961, when Borgward went bankrupt. The Lloyd was good enough to introduce the van concept long before the 1980s – the LT 500 starting in 1953 and the LT 600 since 1955 – combining an engine hood and bus-like design, based on the two-box principle. The high-end 1951 Opel Kapitän was the modern version of a prewar design. It still had long fenders, having rejected the pontoon style, but powerful horizontal chrome bars at the front concealed the formerly high radiator front, and the headlights were recessed into the fenders. But at the top of the list the 1951 Mercedes 300 set new standards, its motto being: We've done it! Its body had the typical radiator, identifiable but barely molded fender contours in front and in back and a very round rear end: a classic, conservative, but appealing car, usually in black, rarely in beige and even more rarely in burgundy. Its later versions as a coupé or cabriolet were rare exemplars of a successful, skillful adaptation of a basic sedan form. It was logical for the then federal chancellor of Germany to choose this car, which is why the 300 was dubbed »Adenauer Benz«. It was very hard for the BMW 501 to compete against it, starting in 1952. Its classic, high kidney shape, its curved, pronounced front and rear fenders and a rounded tail that was formally more emphatic than in the Mercedes 300, its good drivability and sedan-chair comfort were no help: The 501 was doomed forever to be second best among the cars produced during Germany's economic miracle years.

One of the mid-range cars of the economic miracle period, the 1952 Ford 12m had seductive, extremely clearly designed pontoon styling. The sides sported a long straight line stamped into the metal that followed the contours of the front wheels, with a suggestion of rear fins, because the upper edges of the basic body were straight. Small, but very effective was the nose above a two-part, rectangular grille: emblazoned in the middle of it was the globe, a miniature Studebaker motif – a definite modest reference to the U.S. The big Borgward Hansa 2400 did not follow suit. In a bold move, the designer made it a hatchback, unusual in a luxury car – modeled on a design by the aerodynamicist Kamm, who in the 1930s was looking for a compromise between a sensible utilization of space and optimum aerodynamics. A panorama-like windshield with a positively angled A pillar, three side windows, clearly emphasized wheel cutouts, connected with a sill, identical in shape, between them – a feature of the later Isabella – and the large rhombus at the front as a trademark that was recognizable from afar brought no success. Just as helplessly, the follow-up notchback fell short of the 300 or the 501.

In 1953 the GM subsidiary Opel scored with its new Rekord. The basic pontoon form was spruced up with cheek-like bulges over the rear wheels. What was striking, though, was the front of the car, with a wide opening and distinctive ribs on the front grille, like the jaws of a shark. The GM's Chevrolet Corvette had actual teeth. Two things made it world-famous: the synthetic body and elegant European look of a roadster without all the American chrome. In 1953 the good old Mercedes 180 also did without it. The lines of its former fender profiles, stamped in the flanks of the pontoon, had been resurrected decades later in the E-Class series – the misunderstood dic-

48. 1957 Chrysler 300c Cabrio. (Photo: Favcars.)
49. 1955 BMW 507. (Photo: BMW.)
50. 1957 Opel Olympia Rekord. (Photo: Opel.)
51. 1958 Ford Edsel. (Photo: Momentcar.)
52. 1959 Mercedes-Benz 220s. (Photo: Daimler.)
53. 1960 Ford Taunus 17m. (Photo: Spurzem.)
54. 1959 wird der Beweis erbracht, daß ein Ingenieur manchmal der bessere Designer ist. Alec Issigonis konnte auf 305 cm ein Auto mit Platz für vier
Personen und etwas Gepäck gestalten, das gut genug war, auch mal die Rallye Monte-Carlo souverän
zu gewinnen. Heutige Minis sind die größten Lilipu-
taner, wenn ihnen in einigen Varianten 425 cm gut
genug sind. (Photo: Mini.)

es möglich, sportlich, aber preiswert unterwegs zu sein. Der VW Karmann Ghia war mit Käfer-
technik langzeittauglich, seine italienische Formgebung ebenso. Der heutige Blick zurück beweist,
wie ausgewogen der Unterbau mit Kotflügelkonturen das zierliche Dach in richtigem Verhältnis
zur Front- und Heckhaube trägt. Eine Kunst, die die Ghia-Designer bei vielen Studien auf Chrys-
ler-Fahrgestellen in diesen Jahren beherrschten.

Diese Kunst, anders und zierlich zelebriert, hatte 1955 der Deutschamerikaner Graf Albrecht
Goertz mit dem BMW 507 bewiesen. Der Mut, die Niere flach zu legen, den Bug zu pfeilen, die
Räder zu betonen und ein filigranes Hardtop-Dach zu zeichnen, schafft heute siebenstellige Old-
timer-Preise und honoriert so auch die geringe Auflage von insgesamt 254 Exemplaren. Das langt
für die ewig gültige Bewunderung des 507. Ganz anders der Käfer am 5. August desselben Jahres: Es lief der millionste vom Band. Als absoluter automobiler Neustart darf 1955 aber der Citroën DS/ID genannt werden. Alexander Spoerl, damals bekannter Motorjournalist, sagte richtig:
»Der DS ist kein Auto von morgen, sondern von heute. Nur alle anderen sind von gestern.« Recht
hatte er. Das Auto mit der hydropneumatischen Federung konnte notfalls mit drei Rädern fahren.
Viel auffallender aber war seine Architektur: Trennfugen von Hauben und Türen waren logisch verbunden mit Flächenwechseln, Regenrinnen suchten ihr Ende in Blinklichtern, trotz Schrägheck
war der Kofferraum unendlich groß, und die spitze Nase bot dem Wind wenig Widerstand. Innen,
aufgehoben auf weichem Polster, ergänzte das ausgleichende Fahrwerk das Gefühl des Schwe-
bens. Und der Pilot genoß das Sicherheitsgefühl des einspeichigen Lenkrads.

Davon uninspiriert blieben Neulinge in der Zeit danach. Der Opel Rekord von 1957 spielte den
kleinen Amerikaner mit Panoramascheiben an Front und Heck und quälte so das Blech an der
A-Säule, so wie beim 1000 S Coupé von Auto Union 1960 vorne und beim DKW Monza 1956 am
Heck. Geschwungene Chromleisten belebten die Flanken, die sich der Ford Taunus 17m in gezackter Version gönnte, beide Modelle in modischer Zweifarbenlackierung. Dieses Farbenspiel
war dem Winzling Fiat 500 in der Serie verwehrt, nur seine Ableger vieler italienischer Karossiers
gönnten sich das. Oder eben Amerikaner, die auf dem besten Weg waren, die Heckflosse in jene
Höhen zu treiben, die dem Seitenwind gefährliche Angriffsfläche boten. Vertreter dieser Art waren
zum Beispiel der Chrysler 300c von 1957 oder der Cadillac Eldorado von 1959. GM sparte 1959
mit einem Trick: Seine Marken unterschieden sich im Bug und am Heck, das teure Mittelteil mit
Türen war aber baugleich. Ganz ohne Flügel, jedoch mit einer Front, die fachmännische Augen
als Vagina erkennen wollten, fuhr der Ford Edsel 1958 in eine kurze, erfolglose Zukunft. Die aber
hatte der Mercedes 220 S 1959, auch wenn die Stilistiker – der hausinterne Berufsbegriff bei
Daimler – zaghaft Heckflossenansätze formten. Sie nannten sie Peilkanten für Fahrer bei Rück-
wärtsfahrt, meisterten aber nicht den Flächenwechsel von konvex nach konkav. Gemeistert hat
dagegen Sir Alexander Issigonis die Aufgabe, auf 3 m und 5 cm vier Personen unterzubringen.
1959 war dieser Mini, Name ist Programm, der Beweis dafür, daß Ingenieurkunst beste Design-
leistung erbringen kann. Der heutige Mini ist dagegen vergleichsweise nur noch schlicht Designarbeit. Designarbeit vom Besten des ehemaligen Graphikers und danach Ford-Designers Uwe
Bahnsen war 1960 der Taunus 17m. Jede Fläche, jede Kurve und geringer Chrom-Einsatz gaben
dem Werbespruch recht. Es war »die Linie der Vernunft«.

Vorbildliche Designarbeit mit mäßiger Ingenieurleistung bewies 1960 der heckgetriebene
Chevrolet Corvair. Ralph Nader, ein US-Anwalt für Verbraucherschutz, ließ ihn wegen der miserablen Straßenlage vor Gericht sterben. Formal aber war der Corvair Vater vieler Autos, so für den
Fiat 1300/1500 von 1961, den BMW 1500 von 1962, den NSU Prinz mit seinen Ablegern sowie
den Mazda Familia von 1964, den Suzuki 800 fronte und den Panhard 24 bt von 1965, den russischen ZAZ 968 von 1972 und schließlich den Hillman Imp von 1975. Ein anderer Tod traf das

tum of designer-in-chief Bruno Sacco in the 1980s, who had spoken of vertical affinity and horizontal homogeneity, in brief: A Benz must remain recognizable as a Benz. This holds true for the 1954 SL 300: the racing car without a radiator, but sporting a typical star in front and the ever-effective gull-wing doors, independently of many others who used this style of doors. Rightly, world-famous designers chose the SL 300 as the car of the century. A year earlier it became possible to go on the road in a car that was sporty yet inexpensive. The Beetle technology of the VW Karmann Ghia gave it long-term durability, and so did its Italian styling. A look back today shows with what balance the substructure with its fender contours bears the graceful roof in the right proportion to the front and rear hood – an art that the designers of the Ghia had mastered in many study models on Chrysler chassis during these years.

In 1955 the German-American Count Albrecht Goertz had demonstrated this art, in a different form and celebrated for its gracefulness, with the BMW 507. The courage he showed in flattening the kidney grille, in giving the bow an arrow shape, emphasizing the wheels and designing a fili-gree hardtop roof has led to seven-figure Oldtimer awards being given today and thus also honors the small output – a total of only 254 cars. That is enough to ensure that the 507 will forever be admired. Not so the Beetle, on August 5th of the same year: That day, its millionth unit left the production line. The absolutely newest model of 1955, however, was the Citroën DS/ID. Alexander Spoerl, a then-famous automotive journalist, rightly said: »The DS is not the car of tomorrow, but of today. It's just that all the others are cars of yesterday.« He was right, of course. The car with the hydropneumatic suspension could if necessary drive with three wheels. But what was more striking was its architecture: The joints of the hoods and doors were logically connected with level changes, and drip moldings ended in turn-signal lights; in spite of the hatchback the trunk was enormous, and the pointed nose offered little resistance to the wind. Inside, softly cushioned, the balancing chassis and suspension gave passengers the feeling that they were floating. And the driver enjoyed the feeling of safety produced by the one-spoke steering wheel.

The new cars during the period that followed were not inspired by such innovations. The 1957 Opel Rekord acted like a little American car with wrap-around windows in front and rear, thus punishing the panel on the A-pillar; this was also the case in the front of the 1960 Auto Union 1000 S coupé and the rear of the 1956 DKW Monza. Curved chrome strips jazzed up the sides, which in the Ford Taunus 17m were serrated, both models in fashionable two-tone finish. The minuscule Fiat 500 in this series did not come in two colors, only its spin-offs made by many Italian body manufacturers had this feature. Or Americans, who took the tail fin to new heights that made it more vulnerable to crosswind. Among the latter were such cars as the 1957 Chrysler 300c or the 1959 Cadillac Eldorado. GM, in 1959, economized by resorting to a trick: Its brands were differentiated in the front and back, but the expensive midsection with the doors was structurally identical. Completely without wings, but with a front that looked like a vagina to expert eyes, the 1958 Ford Edsel drove off into a brief, unsuccessful future. On the other hand, the 1959 Mercedes 220 S certainly did have a future, even though the style experts – the in-house name Daimler gave them was *Stylistiker* – timidly created rudimentary tail fins. They called them direction-finding ridges for drivers in reverse, but had not mastered the level change from convex to

48. 1957 Chrysler 300c Cabrio. (Photo: Favcars.)
49. 1955 BMW 507. (Photo: BMW.)
50. 1957 Opel Olympia Rekord. (Photo: Opel.)
51. 1958 Ford Edsel. (Photo: Momentcar.)
52. 1959 Mercedes-Benz 220s. (Photo: Daimler.)
53. 1960 Ford Taunus 17m. (Photo: Spurzem.)
54. Here is the proof, in 1959, that an engineer is sometimes better than a professional designer. On a total length of 305 cm Alec Issigonis was able to design a car with room for four people and a bit of luggage that was once good enough to win the Monte Carlo Rally. Today some versions of the largest minis have a total length of as much as 425 cm. (Photo: Mini.)

Haus Borgward 1961. Der Vorwurf an den Chef lautete: Verzettelung mit den drei typenüberfrachteten Marken Borgward, Goliath und Lloyd, letztere speziell mit der flott gezeichneten Arabella und ihrer Sonderkarosserie von Pietro Frua. Schade um die eleganten Autos aus Bremen! Ersatz gab es ausreichend – und neue Marken kamen hinzu. So bewies Renault mit dem R4 im selben Jahr, daß neben der Citroën-»Ente« praktische Autos mit besseren Fahrleistungen möglich waren. Oder daß Engländer sich sportlich aufraffen konnten mit dem Jaguar E-Type, dessen extrem langer Bug samt Kotflügeln geöffnet werden konnte und ein knackiges Heck zeigte, als Coupé-Version wie auch offen. Gar nicht atemberaubend, aber mit logischen Designdetails hatte VW 1961 den VW 1500 über dem Käfer positioniert, ein Stufenheckmodell mit späterer Schrägheck-Variante TL, bösartige Zungen meinten damit die »Traurige Lösung«. Ein Jahr später lockte Opel mit dem neuen Stufenheck-Kadett dem Käfer Kunden ab, was gut gelang. Unaufgeregtes, flächiges Design brachte bequem vier Personen samt Urlaubsgepäck für vier Wochen unter. Leistung und Preis förderten den Erfolg. Japan, weit weg von uns und den USA, begann derweil Autos zu gestalten mit Motiven dicker Amischlitten in halber Größe, aber üppiger Chromausstattung.

Das Ende für die klassische Auslegung der Stufenheck-Limousine mit Frontmotor und Heckantrieb kündigte sich an. Der Mini machte es vor: Steil- oder Schrägheck mit Frontmotor und Frontantrieb beflügelten das Package. Bei gleicher Länge bot dieses Konzept mehr Innenraum, eine ideale Vorgabe für kleine und mittlere Klassen. Der Renault R16 1965 war ebenso Lichtblick wie ein Simca 1100, ein Renault R5 von 1972, ein Audi 50 von 1974 oder eben der Weltbestseller Golf seit 1974. Noch 1967 lästerte der *Spiegel* in Heft 21 über die Modellpolitik mit Hinweis auf die Schelte des damaligen Finanzministers Franz Josef Strauß. Tief getroffen, rückte VW alle noch existierenden Prototypen aus dem »Mausoleum« – so der Name der Modellhalle – auf das Skidpad, die Betonfläche für fahrdynamische Versuche. Weit über 30 Modelle und zum Vergleich beigestellte Konkurrenten wie DKW Junior, BMW 700 oder Glas 1300 bewiesen den VW-Fleiß für Autos mit jeglicher Antriebsart und Motorlage. Erst die finale Entscheidung für das Golfkonzept mit Hilfe des jungen Designers Giorgetto Giugiaro und hausinterne Nachbesserungen (z. B. anstelle eckiger Scheinwerfer runde – es sind ja die »Augen« eines Autos – und Verzicht auf sichtbare Entlüftungsschlitze der Kabine) wurden zur Basis für den anfänglich rostigen, später aber beständigen Erfolg des Golfs und des baugleichen Coupés Scirocco, das schon im Frühjahr 1974 der Presse wegen Ölkrise und Sonntagsfahrverbot auf dem eigenen Testgelände vorgestellt wurde, der Golf dagegen erst am herbstlichen Nymphenburger Schloß. In Zürich begrüßte der Technikableger Passat die Presse, während der Audi 50 als weiterer Ableger die Presse auf Sardinien empfing.

Für die Welt der Reichen wurde 1966 der Miura geboren. Der Weinbauer und Treckerproduzent Lamborghini schuf dieses Edeltier (Miura ist eine Stierrasse). Wahrscheinlich ist der Miura der Vater aller wabenförmigen Grillgitter dieser Welt, das bei ihm am Bug ganz weit unten sitzt. Das Design stammt vom Bertone-Mitarbeiter Marcello Gandini. Heute ist VW Eigner von Lamborghini, die bis heute variantenreich extrem flache »Flundern« bauen. Nicht flach, aber langhaubig wie ein E-Type präsentierte sich 1964 der Ford Mustang, ein Auto mit optisch sportlichen Ambitionen, das nach Jahren verfetteter Karosserien wieder die Ursprünge belebte. GM hielt dagegen mit dem Chevrolet Camaro und dem Pontiac Firebird. Deutschland folgte diesem Trend erfolgreich

55. 1966 wird der Traum aller Sportwagen Realität: der Lamborghini Miura aus der Hand von Bertone sieht schnell aus und ist es. (Photo: Lamborghini.)
56. 1964 Opel Kapitän A wird die preiswerte Alternative zum Mercedes 220S. Schon behaupteten Fachkundige: das beste deutsche Auto wäre ein Mercedes mit Opelmotor. (Photo: Opel.)
57. 1966 kann die DDR zeigen, daß ein Trabi nicht alles ist. Der Wartburg 353 ähnelte in der Haubenöffnung an Bug und Heck dem BMW 1500 und hatte an seinem Ende gar einen Viertaktmotor. (Photo: Wartburg.)
58. 1967 kann der Toyota Corona 1900 auf Pontonbasis Ähnlichkeiten zur Seitenansicht des Fiat 124 nicht verleugnen. (Photo: Toyota.)
59. 1966 macht der Renault R16 klar, daß es auf knapper Länge mehr Raum gibt als bei Stufenheckversionen. Dank Schrägheck kann der Innenraum extrem flexibel ausgelegt werden. (Photo: Wikipedia.)
60. 1970 zeigt der Opel Manta, daß amerikanische Designideen zivilisiert werden können. Weniger Show als bei einem Ford Capri, mehr Klarheit an Bug und Heck. (Photo: Opel.)
61. 1974 geht in Wolfsburg eine neue Sonne auf. Kein Käfer mehr, jetzt Frontmotor mit Frontantrieb bei 30 cm weniger Länge: der Golf auf dem Weg zum späteren Produktionsweltmeister. (Photo: Volkswagen AG.)

55. In 1966 the dream of all sports cars becomes reality: Designed by Bertone, the Lamborghini Miura not only looks fast, it is fast. (Photo: Lamborghini.)

56. The 1964 Opel Kapitän A becomes the low-cost alternative of the Mercedes 220S. Experts claimed that the best German car was a Mercedes with an Opel engine. (Photo: Opel.)

57. In 1966 the GDR is able to demonstrate that it can produce more than Trabis. The hood opening of the Wartburg 353 in front and in the rear resembled that of the BMW 1500, and towards the end the Wartburg even had a four-stroke engine. (Photo: Wartburg.)

58. The 1967 Toyota Corona 1900 with a pontoon base has undeniable similarities with the side view of the Fiat 124. (Photo: Toyota.)

59. The 1966 Renault R16 clearly demonstrates that with a shorter length there is more room than in pontoon-type versions. Thanks to the hatchback the interior space can be laid out extremely flexibly. (Photo: Wikipedia.)

60. In 1970, the Opel Manta shows that American design ideas can lead to civilized results. Less showy than a Ford Capri, with more definition in the front and rear section. (Photo: Opel.)

61. In 1974 a new era begins in Wolfsburg. No longer a beetle, but 30 cm shorter, with a front engine and front wheel drive, the Golf is on the way to breaking the world car production record. (Photo: Volkswagen AG.)

concave. On the other hand, Sir Alexander Issigonis mastered the task of accommodating four persons on 3 m and 5 cm. In 1959 this mini – the name says it all – was proof that the art of engineering is able to produce the best designs. Today, on the other hand, the mini in comparison is mere design work. In 1960, the Taunus 17m was the top-of-the-line work of the former graphic designer and later Ford designer Uwe Bahnsen. Every surface, every curve, and the sparing use of chrome, proved the advertising slogan right. It was »the line of good sense«.

Exemplary design work with moderate engineering performance was demonstrated in 1960 by the rear-wheel-drive Chevrolet Corvair. Ralph Nader, an American attorney involved in consumer protection, caused its demise in court because of the miserable way it held the road. But from a formal point of view the Corvair sired many cars, for instance, the 1961 Fiat 1300/1500, the 1962 BMW 1500, the NSU Prinz with its spin-offs and the 1964 Mazda Familia, the Suzuki 800 fronte and the 1965 Panhard 24 bt 1965, the 1972 Russian ZAZ 968 and finally the 1975 Hillman Imp. A different kind of death was in store for the Borgward company in 1961. The head of the firm was accused of wasting company money on the three brands Borgward, Goliath und Lloyd, which were overloaded with models – in the case of Lloyd in particular with the snazzily designed Arabella and its custom-made body by Pietro Frua. What a shame about the elegant cars from Bremen! There were plenty of others to take their place – and new brands were certainly added. Thus in the same year Renault with its R4 proved that in addition to the Citroën »Duck«, practical cars with better driving performance were actually possible. Or that the British could create a better sports car with the Jaguar E-Type, whose extremely long front along with the fenders could be opened and showed a sexy rear section, in both the coupé version and the open version. Hardly breathtaking, but featuring logical design details, the 1961 VW had positioned the VW 1500 above the Beetle, a pontoon-type model with a later hatchback variant TL, which malicious tongues interpreted as the »Traurige Lösung«, or sad solution. One year later Opel with its new pontoon model Kadett lured away the Beetle's customers without too much difficulty. Its unexciting, two-dimensional styling comfortably accommodated four persons plus vacation luggage for four weeks. Performance and price contributed to its success. In the meantime Japan, far away from Germany and the U.S., began designing cars that looked like the bulky Yankee muscle cars but were half the size and had plenty of chrome.

The end of the classic interpretation of the saloon car with a front engine and rear drive was in sight. The mini was the pioneer: A hatchback or coupé back with a front engine and front drive enhanced the package. With the same length, this design provided more interior space, an ideal feature for small and mid-size cars. The 1965 Renault R16 was a bright spot, and so were a Simca 1100, a 1972 Renault R5, a 1974 Audi 50 or the world bestseller Golf, starting in 1974. As late as 1967, issue no. 21 of *Spiegel* magazine ranted about model policy with reference to the criticism of then German Minister of Finance Franz Josef Strauss. Cut to the quick, VW moved all still existing prototypes out of the »Mausoleum« – as VW called its hall of models – onto the skidpad, the concrete area used for testing car performance. More than 30 models, and competitors made available for comparison such as the DKW Junior, BMW 700 or Glas 1300, demonstrated VW's desire to design cars with every possible type of drive and motor position. Only the final decision in favor of the Golf design with the help of the young designer Giorgetto Giugiaro and subsequent in-house improvements (e.g., round headlights in lieu of square ones – for they are the »eyes« of a car – and no visible ventilation slits in the cab) became the basis for the initially sluggish, but later enduring success of the Golf and the structurally identical coupé Scirocco, which in the spring of 1974 was introduced to the press on the company's own test track due to the oil crisis and the ban on Sunday driving, while the Golf was not shown until that fall at the Nymphenburg Palace. In Zurich its mechanical spin-off, the Passat, greeted the media, while another spin-off, the Audi 50, was shown off to the press in Sardinia.

The year 1966 saw the birth of the Miura, intended for wealthy customers. The winegrower and tractor manufacturer Lamborghini created this noble beast (Miura is a breed of bulls). In all likelihood the Miura is the father of all honeycomb-like grills in the world: In the Miura, the grill is way low on the front. The design was the creation of Marcello Gandini, who worked for the Bertone design group. Today, VW owns Lamborghini, which to this day has been building a large variety of extremely flat »flounders«. Not flat, but long-hooded like an E-Type was the 1964 Ford Mustang, a car with the ambition of achieving a sporty look, which after years of having potbellied bodywork returned to its sleek origins. Meanwhile GM stood by the Chevrolet Camaro and the Pontiac Firebird. Germany successfully followed the trend in 1969 with the Ford Capri; in 1970 Opel came up with the Manta, which was the butt of many jokes but had an unpretentious overall shape, while the Capri laid it on thick. A far more serious car, but universally loved, was the 1968

1969 mit dem Ford Capri und Opel 1970 mit dem Manta, der Vater vieler Witze wurde, dafür eine unprätentiöse Gesamtform bot, während die Capri etwas dicker auftrug. Ohne hämische Witze, aber 1968 von allen geliebt, war der Opel GT als Corvette des kleinen Mannes. Italienisch zierlich und bescheiden klein zeigte sich dagegen der Welt erstes Auto mit Wankelmotor, der 1965 NSU Wankel Spider. Die dachlose Version des Sportprinzen mit versenkbarem Stoffdach oder wahlweise montierbarem Hardtop war vorne fast ein Alfa und hinten zart heckflossenbeflügelt, geformt von Bertone und insgesamt 2375mal gebaut. Aus eigenem Hause stammte 1967 der Entwurf für den NSU Ro 80. Der Karosserieingenieur Claus Luthe mit Sinn für Autodesign schuf dank des flach bauenden Motors ein Auto mit formaler Weltgeltung. Das Prinzip flache Front und hohes Heck wurde perfekt gelöst und ist heute Standard der Stufenheckautos. Das kleinere Modell von 1970 mit herkömmlichen Hubmotor war dagegen ein praktischer, kantiger Entwurf und wurde nach dem Verkauf von NSU an VW ebenso schlicht getauft: K 70.

Schlicht machten es auch die USA. Der Druck, kleinere Autos anzubieten, war groß. Selbst Cadillac machte mit. Seine Version von 1980 als Seville kopierte zugleich die englische »Razoredge«-Mode: rasiermesserscharfes Blech in seinen Formübergängen, edel genug, um damit bereits 1949 einen kleinen Triumph Mayflower zu beglücken. Nicht beglückt war die Autowelt, als Alfa Romeo aus hilfloser Modellpolitik 1983 aus einem Nissan Cherry einen Alfa Arna machte, der beste Weg, sein Image zu töten. Da hatte Fiat mit dem Uno aus der Feder von Giugiaro 1983 mehr Fortune: klares Konzept, klare Glas- und Blechflächen, zwei- und viertürig. Daß es mit dieser Designeinstellung auch kleiner ging, bewies der Renault Twingo 1992: das perfekte One-Box-System und der fröhlichste Minivan weit und breit. Im selben Jahr brachte Mercedes den »Baby-Benz« auf den Markt. Geschickt wurde die Mittelklasse bei dem 190 praktisch auf 0,8-Größe verkleinert und damit seine kompakte Form kaschiert, die erst im Vergleich mit üblichen Modellen ihre wahre Größe zeigte. Unglücklich begann der Start des kleinsten Mercedes, eines One-Box-Autos mit ungewohnter Fenstergestaltung – jemand nannte es »deutsches Fachwerk«. Bei der Pressepräsentation legte 1997 ein schwedischer Journalist während der Lastwechselfahrt durch die Poller das Auto aufs Dach. Mehr als 4000 bereits ausgelieferte Autos wurden kostenlos gegen nachgebesserte Modelle ausgetauscht, seither heißt das Malheur »Elchtest«. Das Konkurrenzmodell Audi A2 von 2001 machte es besser. Mit aufwendiger Aluminiumkarosserie mit Bestwerten im Windkanal und ungewöhnlicher Seitenansicht als Hommage an die Kamm-Form war das Auto auf Dauer zu teuer und wurde bald vom Markt genommen – heute ist es eine gesuchte Rarität. Ein ähnliches Schicksal hatte der Renault Avantime von 2001: Ein ungewöhnliches Van-Design mit übergroßen Türen und formbestimmender Zweifarbenlackierung betonte den neuen Designstil vom Chefdesigner Patrick Le Quément. Der große Vel Satis und der Megane mit steilem Heckfenster blieben ähnlich glücklos: Das MAYA-Prinzip läßt grüßen!

Inzwischen reussierte die automobile Vergangenheit: Wo Japan mit einem Daihatsu Copen 2003 in Europa oder bereits 1967 der Fiat Vignale Gamine und der Fiat 500 von 2007 das Retrodesign liebten, da zog VW 1997 mit. Sein New Beetle war designmäßig die große Kunst der Radien und interpretierte geschickt die klassische Käferform, alles auf den Rädern des Golf. Sein Nachfolger streckte sich mehr und verwässerte den Urkäfer-Ansatz. 2004 begründete Mercedes mit dem CLS die Mode der coupéhaften flachen Limousinen, hier fließend weich, in der Shooting-Brake-Version aber leicht überzogen. Audi zeigte sich 2008 mit den Schrägheckvarianten A5 und A7 dagegen konturierter, ohne Eleganz zu verlieren. Daneben geriet diese Idee, weil zu ruppig, bei BMW mit den Versionen X6 von 2008 und X4 von 2014. Nicht ruppig, aber unglücklich war das Design des Panamera, der ab 2009 als viertüriger Porsche die Luxusklasse der Limousinen sportlich aufmischte. Kritisiert wurde der bucklige Dachabschluß im Heck, ein Fehler, den das

62. 1968 Opels GT-Werbung hatte recht: Nur fliegen ist schöner, noch schöner die Form. (Photo: Opel.)

63. 1965 wird der NSU Wankel-Spider der Welt erstes Auto mit Wankelmotor, und das 2375mal. (Photo: Autor)

64. 1980 zeigt uns der Cadillac Seville, daß man englische Karosseriekunst kopieren kann, nur der britische Charme fehlt. (Photo: GM.)

65. 1983 killt der Alfa Romeo Arna seine Marke, denn die Technik stammt vom japanischen Aschenputtel Nissan Cherry. (Photo: Alfa-Romeo.)

66. 1980 kann der Fiat Panda dank seiner kargen, aber sinnvollen Ausstattung die Werber inspirieren, die ihn die »tolle Kiste« nannten. Hier kann Giugiaro mit geringsten Mitteln die sinnvollsten Funktionszusammenhänge bündeln. (Photo: Fiat.)

67. 1983 beweist Giugiaro, daß sein Fiat Uno (hier Version 55S) mit konservativer Blechtechnik optimalen Nutzwert und Langzeitschönheit bieten kann. Dabei hilft das Technikkonzept Frontmotor, Frontantrieb und Steilheck, ein Rezept wie beim Golf, der auch von Giugiaro ist. (Photo: Fiat.)

68. 1992 lernen wir am Renault Twingo, daß die klare Ratio etwas Charme verträgt. Mit nur wenigen Details wie den Scheinwerfern ist das machbar und versteckt so die gehörige Portion Vernunft dahinter. (Photo: Renault.)

Opel GT, the Corvette of the little man. On the other hand, the world's first car with a Wankel engine, the 1965 NSU Wankel Spider, looked gracefully Italian and modestly small. This roofless version of the sport prince, with a retractable fabric roof or optional mountable hardtop, was almost an Alfa in front and had delicate tail fins; the body was by Bertone. A total of 2,375 Wankel Spiders were built. The design for the 1967 NSU Ro 80 was the company's own. Thanks to the slim-line engine, the talented automotive body engineer Claus Luthe created a car of formal international standing. It solved the problem of combining a flat front and high rear perfectly, and is today the norm for notchback cars. On the other hand, the smaller model, built in 1970, which had a conventional hoist motor, was a practical, angular design and, after NSU was sold to VW, was given the equally plain name of K 70.

Plain and simple was a trend in the U.S. as well. There was great pressure to offer the public smaller cars. Even Cadillac joined in. Its 1980 version – the Seville –at the same time copied the British »Razoredge« fashion: a sharp-edged body in its form transitions, classy enough to be used in 1949 on the little Triumph Mayflower. But the automotive world was not as delighted when in 1983 Alfa Romeo due to its helpless model policy made a Nissan Cherry into an Alfa Arna, the best way to ruin its image. Fiat was luckier with its 1983 Uno designed by Giugiaro: a clear concept, clear glass and metal surfaces, two- and four-door. This design configuration could be kept smaller as well, as demonstrated by the 1992 Renault Twingo: the perfect one-box system and the jauntiest minivan far and wide. In the same year Mercedes brought the »Baby Benz« on the market. In the Mercedes-Benz 190, the mid-size class was cleverly reduced by virtually one-fifth, thus concealing its compact form, whose true size became apparent only when compared to conventional models. The smallest Mercedes, a one-box car with an unusual window design – someone called it »German half-timbering« – had an unlucky start. When it was presented to the press in 1997, during the load-reversal test drive through the bollards, a Swedish journalist flipped the car on its roof. More than 4000 already delivered cars were exchanged free of charge for improved models – ever since, the disastrous procedure has been called the »elk test«. The competing model, the 2001 Audi A2, outdid the Mercedes. With a costly aluminum body, optimum values in the wind tunnel and an unusual profile – an homage to the K-form – the car proved to be too expensive in the long run and was soon taken off the market – today it is a sought-after rarity. The 2001 Renault Avantime suffered a similar fate: An unusual van design with oversize doors and two-tone paintwork that determined its form accentuated the design style of designer-in-chief Patrick Le Quément. The large Vel Satis and the Megane with its steep rear window were to be similarly unlucky: Remember the MAYA principle!

In the meantime, the automotive past was successful: While Japan, with a 2003 Daihatsu Copen in Europe, or even the 1967 Fiat Vignale Gamine and the 2007 Fiat 500 loved retro design, VW went along with the trend in 1997. In terms of design, its New Beetle represented the great art of working with circles and cleverly interpreted the classic beetle shape, all on the wheels of the Golf. Its successor was more elongated, and watered down the original Beetle concept. With the 2004 CLS, Mercedes established the fashion of coupé-like flat sedans, soft and flowing in this version while slightly exaggerated in the shooting-brake version. The 2008 Audi fastback variants, the A5 and A7, on the other hand, were more contoured, while retaining their elegance. This idea failed in the case of BMW, being too rough-looking, with the versions X6 (2008) and X4 (2014).

62. The advertisement for the 1968 Opel GT was right: Only flying is better, and the form is even greater. (Photo: Opel.)
63. In 1965 the NSU Wankel Spider becomes the first car in the world to have a Wankel engine, multiplied by 2375. (Photo: author's archive.)
64. The bodywork of the 1980 Cadillac Seville proves that you can certainly copy British craftsmanship – only the British charm is missing. (Photo: GM.)
65. In 1983 the Alfa Romeo Arna kills its brand, for its technology originates with the Japanese Cinderella Nissan Cherry. (Photo: Alfa Romeo.)
66. The sparse but effective features of the 1980 Fiat Panda inspire advertisers, who call it an »awesome box«. Here, with the simplest of means, Giugiaro is able to combine the functional correlations that make the most sense. (Photo: Fiat.)
67. Giugiaro proves, in 1983, that with conservative metal technology his Fiat Uno (pictured is the 55S version) can offer optimum utility and long-term great looks. Another plus is the technical concept of having a front engine, front-wheel drive and hatchback, a recipe that works for the Golf, another Giugiaro design. (Photo: Fiat.)
68. What we learn from the 1992 Renault Twingo is that clear rational logic could do with a bit of added charm. This can be achieved by means of just a few details, such as the headlights, thus concealing the rational thinking behind the design. (Photo: Renault.)

69. 2006 kann VW mit dem Beetle dem Trend des Retrodesigns folgen – mit dem Golf als Basis. (Photo: Volkswagen AG.)

70. 1982 bietet Mercedes für das kleine Portemonnaie den 190, auch Baby-Benz genannt. (Photo: Daimler.)

71. 1997 macht die Mercedes A-Klasse Schlagzeilen: Es schlug sie aus der Bahn, was ihre praktische Länge von 370 cm nicht störte. (Photo: Daimler.)

72. 2001 wollte der Audi A2 mit der A-Klasse konkurrieren, war aber mit Alu-Karosse auf Dauer zu kostspielig. (Photo: Audi.)

73. 2008 macht der Audi A5 klar, daß die Marke mit der Mode gehen kann. Eine Limousine mutiert zum viertürigen Coupé. (Photo: Audi.)

74. 2007 zeigte BMW den X6 Concept und bewies, daß nicht alles aus dem Hause elegante Volumen haben kann. (Photo: BMW.)

75. 2011 kam der Opel Ampera elektrisch zu früh, es fehlte die Steckdoseninfrastruktur. (Photo: Opel.)

76. 2013 konnte man für 110.000 Euro den VW XL1 als Weltmeister des Cw-Wertes in die Garage stellen, aber nur 200 gingen in den Handel. (Photo: Volkswagen AG.)

77. 2017 will der Land-Rover Discovery zeigen, daß man optische Muskeln auch vornehm verpacken kann. (Photo: Land-Rover.)

Nachfolgemodell perfekt retuschierte und dieses Auto jetzt augenfreundlich zeigt wie seinen Shooting-Brake-Ableger Panamera Sport Turismo von 2016.

Was die Welt umtreibt, ist das CO_2-Problem, bei dem Automobile beteiligt sind. Seit langem werden alternative Konzepte erforscht. Zwischenschritte wie Hybridantrieb oder Brennstoffzellenmotore suchen alle den abgasfreien Weg, ausgebremst durch den VW-Dieselskandal von 2015 und viele Marken mit überhöhten Abgaswerten. Toyota bemüht sich um saubere Luft seit 1997 mit dem Schräghheckhybriden Prius, Opel von 2012 bis 2016 mit dem Ampera von GM als Hybrid oder mit Batterie ohne großen Erfolg, aber mit auffallender Lichtgraphik im Bug, geliehen vom Van Zafira, dessen Innenraumvariabilität gelobt wurde. Frei von Konzerngremien, weckte die Marke Tesla aus den USA 2008 die Branche mit ihrem Roadster auf. Als Basis diente der Lotus Elise für den reinen Elektroantrieb. Bald wurden weitere Modelle angekündigt – und tatsächlich auch gebaut. Als große Limousine Tesla S seit 2012 überflügelte das Modell deutschen Edelstahl wie die Mercedes S-Klasse oder den 7er BMW. Der SUV Tesla X von 2016 bedient Anhänger dieser Autogattung elektrisch. Alle Tesla zeigen in gesunder Mischung klares Design, das dem MAYA-Prinzip gehorcht. VW versuchte sich anders und war 2013 mit dem XL1 nicht nur extrem windschnittig (Cw-Wert 0,186), sondern kam als Hybrid mit einem Liter auf 100 Kilometer aus. 200 Exemplare und ein Verkaufspreis von 111 000 Euro sind die Basis für jene, die künftig einen klassischen Oldtimer suchen. Praktischer war es, 2013 den VW e-Up zu wählen, der, seit 2011 benzinbetrieben, mit eigener Formensprache den Markt der Kleinstwagen auch elektrisch beförderte, so wie die Konkurrenz von Mitsubishi seit 2009 mit dem MiEV in trostlosem Karosseriekleid fährt, das auch die Zwillinge Citroën C-Zero und der Peugeot iOn tragen. Der Renault Zoe ist seit 2012 in beschwingt seriöser französischer Handschrift besser unterwegs. Die mit chinesischem Kapital wiederbelebte Marke Borgward hofft seit 2015, mit dem elektrischen SUV BX den Anschluß mit einem Design zu finden, das brav, aber ohne eigenen Charakter ist. Den dürfen wir zum Beispiel von den neuen VW-Studien erwarten, die 2016 vorgestellt wurden. Ein Teil der Autoindustrie vertraut weiterhin auf die Zukunft des Benzins. Und ausgerechnet die sportlichen Marken wie Jaguar, Alfa Romeo oder Maserati beglücken uns mit SUVs, die nur zu gerne die 200-km/h-Marke knacken. Da loben wir den Land Rover, der in seinem bisherigen Leben formale Ruhe in allen Versionen ausstrahlt. Alle anderen PS-starken SUVs, ob klein oder groß, machen weiter mit in diesem Wettrennen, auch in der formalen Übertreibung im Design.

Not rough-looking but unfortunate was the design of the Panamera, a four-door sporty-looking Porsche that from 2009 on stirred up the sedan luxury class. Critics found fault with the hunchback roof termination in the rear, an error that the follow-up model perfectly amended, giving this car eye appeal, like its shooting-brake spin-off, the 2016 Panamera Sport Turismo.

What worries the planet is the CO_2 problem, and that involves automobiles. Scientists have been searching for alternative concepts for a long time. Intermediate solutions like the hybrid drive or fuel-cell engines are all looking for an emission-free fix, thwarted by the 2015 VW diesel scandal, and many brands with excessive exhaust emission values. Toyota has been striving for clean air since 1997 with the hatchback hybrid Prius, while Opel from 2012 until 2016 manufactured the GM Ampera as a hybrid or with a battery, without much success but with striking light graphics in the front, borrowed from the van Zafira, which was praised for its interior flexibility. In 2008 the American brand Tesla, which did not have to worry about boards of directors, roused the industry with its roadster. It was based on the Lotus Elise for its all-electric drive. Soon additional models were announced – and actually built. As the large sedan Tesla S, first manufactured in 2012, this model outstripped such German high-grade cars as the Mercedes S-Class or the BMW 7 series. The 2016 SUV Tesla model X is all-electric. All Tesla models are a healthy mix of clear design that follows the MAYA principle. VW had a go at doing it differently, and its 2013 XL1 was not only extremely streamlined (drag coefficient: 0.186), but as a hybrid also managed with one liter/100 kilometers. Two hundred XL1s at a price of 111 000 euros are out there for those who will be looking for a classic old-timer in future. It was more practical to choose the VW e-Up in 2013: gasoline-powered since 2011 and with unique styling, this subcompact also comes in an electric version, just as the competitor of Mitsubishi has, since 2009, been building the MiEV whose dreary bodywork it shares with the twins Citroën C-Zero and the Peugeot iOn. Since 2012, the Renault Zoe, which bears a vibrantly serious French signature, has been in a better position. Since 2015 the Borgward brand, revitalized with Chinese capital, has been hoping to catch up with its competition with the electric SUV BX, a decent design, but without a unique character of its own. Character is something we can expect, say, from the new cars developed by VW that were introduced in 2016. A section of the auto industry continues to have confidence in the future of gasoline. And it is sport cars like the Jaguar, Alfa Romeo or Maserati that delight us with SUVs which like nothing better than to break the 200-km/h mark. That's why we commend the Land Rover, all of whose versions to date have been formally humdrum. All the other powerful SUVs, whether small or large, continue to compete in this race, including formal exaggeration when it comes to design.

69. In 2006, with the Beetle, VW follows the trend of retro design – with the Golf as a platform. (Photo: Volkswagen AG.)

70. In 1982 Mercedes offers its lower-priced 190, aka Baby Benz. (Photo: Daimler.)

71. The Mercedes 1997 A-Class hits the headlines: It's off the beaten track, a fact undiminished by its practical length of 370 cm. (Photo: Daimler.)

72. The 2001 Audi A2 tried to compete with the A-Class, but in the long term its aluminum body was too expensive. (Photo: Audi.)

73. The 2008 Audi A5 brings home the fact that the brand can go along with fashion. A sedan has mutated into a four-door coupé. (Photo: Audi.)

74. In 2007 BMW revealed the X6 Concept, proving that not all their vehicles have elegant volumes. (Photo: BMW.)

75. In 2011 the electric car Opel Ampera arrives too soon, as there is no plug-in infrastructure. (Photo: Opel.)

76. In 2013, for 110 000 euros, you could have a VW XL1 – a world champion when it comes to drag coefficient – in your garage, but only 200 were sold to retail customers. (Photo: Volkswagen AG.)

77. The 2017 Land Rover Discovery demonstrates that optical muscles can come in an elegant package. (Photo: Land Rover.)

4. Verrückte Technik schafft skurrile Form

So ist das mit Erfindungen: Klappt das Prinzip, suchen schlaue Köpfe gleich nach Alternativen. Das Auto, gerade in seinem Grundkonzept festgeschrieben mit Frontmotor, vier Rädern und Kardanwelle, die bald den Riemen- oder Kettenantrieb ablöste, machte Mutige mutiger und Übermutige zu waghalsig. Vermutlich inspiriert von der fast zeitgleichen Entwicklung fliegender Kisten, anders sind die Anfänge der Fliegerei nicht zu nennen, suchte man den leichteren Weg des Vorankommens. Der Motor als Grundvoraussetzung blieb für den Vortrieb. Warum aber bedarf es da noch eines Riemens, einer Kette, einer Kardanwelle, eines Differentials, wenn es auch viel einfacher geht? Da langt zum Fahren doch ein Propeller am Bug oder am Heck, nur die Bremsen mußten bleiben. Ein Rumäne, leider nicht überliefert ist das Jahr seines Übermuts, ging einen Schritt weiter und hatte vermutlich das erste Automobil-Flugzeug gebaut. Seine Flügel erinnerten an Lindberghs Gleitsegler oder andere erste Flugmaschinen, die schiere Funktion zeigten. Bänderverspannte Tragflächen im Wortsinn hatten nichts gemein mit unserer Vorstellung von einem üblichen Flugzeug. Das Fahrwerk war hier eine Vergrößerung in Richtung automobiler Ähnlichkeit. Ob dieser Apparat jemals abhob oder gar regelrecht fliegen konnte, ist nicht überliefert und aus dem Photodokument nicht ersichtlich.

Auf dem Boden der Tatsachen blieb jener Bastler, der sich den hölzernen Pferdewagen auslieh, die Deichseln absägte, einen Tank hinter dem Fahrerrücken auf Füße stellte und vor sich einen veritablen V-Motor mit wuchtigem Kühler montierte. Das Entsetzliche aber war ein Propeller mit gut 3 m Durchmesser. Man stelle sich vor, daß dieses ungelenke Gefährt auf schneller Fahrt nur durch primitivste Bremskonstruktion zum Halten zu bringen war, wobei die Bremsweglänge unerwähnt blieb. Indirekt heißt das: Hier hat ein mutiger Tüftler lediglich das Vorankommen gefördert, wenig, fast gar nicht die Gefahr der kraftvollen Rotation des Monster-Propellers berücksichtigt. Der verhalf eher dem Fahrer und eventuell direkt nachfolgenden Radfahrern zu einem windigen Ausritt, was dem Begriff Fahrtwind eine neue Bedeutung schenkte. Dankbar muß man sein, daß diese Grille aus dem Jahre 1922, glaubt man dem Datum auf der Photographie, die automobile Geschichte nicht bereicherte und nicht weiter von sich hören ließ.

Mehr Rücksicht auf andere Verkehrsteilnehmer, soweit sie vorwegfuhren, zeigte der Curtiss Autoplane von 1917. Einerseits veränderte sich die Form der Flugmaschine zu einem erkennbaren Flugzeug – hier in der Doppeldeckerversion mit dritten Stummelflügeln –, andererseits verband der Karosseriekörper Flugzeugrumpf und Autokarosserie in akzeptabler Kombination. Und erste Erkenntnisse aerodynamischer Gestaltungsprinzipien spiegelten sich in der geneigten Frontscheibe und dem leicht gerundeten Rumpf wider. Weil aber der Propeller am Heck zum Glück etwas höher saß, ärgerten sich nachfolgende Radfahrer über Gegenwind. Der verflüchtigte sich schneller bei allen Konstruktionen des Franzosen Marcel Leyat, der zwischen 1922 und 1932 diverse Varianten des vorne eingebauten Propellers in einer Kleinserie baute und unbeirrt an diese Technik für die Zukunft glaubte. Mal war die Karosserie geschlossen, einem Flugzeugrumpf ähnelnd, mal ein offenes Gefährt. Da der große Propeller nur notdürftig mit etwas Draht vor Eingriffen geschützt war, konnten die Vorderräder nicht mit einer Achsschenkellenkung versehen werden, die eingeschlagenen Räder hätten sonst den Rahmen des Propellers berührt. So mußte eine Hinterradlenkung aushelfen. Wer weiß, wie schwierig hier manövriert werden kann, sollte sich

78. Aller Anfang ist schwer auf dem Weg in die Luft – auch in Rumänien. (Photo: Internet.)
79. 1917 macht der Curtiss Autoplan eine fast gute Autofigur mit Flügeln. (Photo: Internet.)
80. Nicht fliegen, nur vorwärts wollte er – aber so umständlich. (Photo: Internet.)
81. 1946 ging ein komplettes Auto in die Luft: der Convair Model 118 von Theodor Parsons Hall. (Photo: ETS Quebec.)
82. 1932 machte der französische Helicron No.1 den Versuch, einen lebensgefährlichen Antrieb für gut zu befinden. (Photo: GT-Planet.)
83. 1949 begann der Versuch, so etwas wie das Molt-Taylor-Aerocar in die Luft zu bringen. (Photo: R. Carlson.)
84. 1968 meint die brasilianische Uni FEI, daß ihr X1 etwas Sinnvolles sei. (Photo: FEI.)

78. Every beginning is difficult as you ascend in the air – even in Romania. (Photo: Internet.)
79. In 1917 the Curtiss Autoplane looked pretty good for a car with wings. (Photo: Internet.)
80. He didn't want to fly, just move forward – but it was ever so involved! (Photo: Internet.)
81. In 1946 a complete car went up in the air: the Convair Model 118 by Theodor Parsons Hall. (Photo: ETS Quebec.)
82. The French Helicron No.1, in 1932, tried to demonstrate that a life-threatening drive mechanism was okay. (Photo: GT-Planet.)
83. 1949 marked the beginning of the experiment to fly something similar to the Molt Taylor Aerocar. (Photo: R. Carlson.)
84. In 1968, the Brazilian Central University of FEI thought its X1 made sense. (Photo: FEI.)

4. Crazy technology leads to bizarre form

That's how it is with inventions: If the principle works, clever brains immediately start looking for alternatives. The car, whose basic concept was just laid down as having a front engine, four wheels and a drive shaft that soon replaced the belt or chain drive, made bold people bolder and overconfident people too daring. Presumably inspired by the almost simultaneous development of flying crates – there is no other name for the beginnings of aviation – people looked for the easier way of locomotion. The engine – a prerequisite – was still used for propulsion. But why have a belt, a chain, a drive shaft, a differential, if there's also a much easier way? In order to drive, all you need is a propeller, in front or in the rear, only the brakes had to stay. A Romanian – unfortunately the year he did the reckless deed is not recorded – went a step further and presumably built the first automobile airplane. His wings were reminiscent of Lindbergh's hang-glider or other early flying machines that were examples of pure function. Wings braced with straps had nothing in common with our idea of a conventional plane. This vehicle already looked more like an enlarged automobile. It is not recorded, nor obvious from the photo document, whether this device ever lifted off or was even capable of regular flight.

One down-to-earth do-it-yourselfer took a wooden carriage, sawed off the shafts, put a tank behind the driver's back and mounted a veritable V-type engine with a hefty radiator in front of himself. But what was horrific was a propeller that was a good 3 m in diameter. Imagine this clumsy vehicle at high speed: It could be brought to a stop only by the most primitive braking system, and there was no mention of the braking distance. Indirectly this means: Here, a courageous tinkerer had been working exclusively on locomotion, paying little or almost no attention to the danger posed by the powerful rotation of the monster propeller. That propeller did help the driver and any cyclists following directly behind him to have a breezy ride, giving the term »head wind« a new meaning. We should be thankful that this crazy notion back in the year 1922, if we are to believe the date on the photograph, never took off and was never heard of again.

The 1917 Curtiss Autoplane was more considerate of other people who used the road, as long as they were riding ahead of it. On the one hand, the form of the flying machine had changed into a recognizable aircraft – shown here in the double-decker version with added canard wings; on the other hand, its body had both an aircraft fuselage and an automotive body in acceptable combination. And early insights into aerodynamic design principles were reflected in the sloping windshield and the slightly rounded fuselage. But because the propeller was located in the rear, and luckily somewhat higher up, cyclists who followed the vehicle were annoyed by the head wind. There was less headwind in all the designs of the French inventor Marcel Leyat, who, between 1922 and 1932, built a number of variants of the front-mounted propeller in a small series and firmly believed in the future of this technology. In some of his models the body was closed, resembling the fuselage of an aircraft, in others the vehicle was open. Since the large propeller was only barely protected from interference with a bit of wire, the front wheels could not have Ackerman steering, since the covered wheels would otherwise have touched the frame of the propeller. That is why rear-wheel steering had to be used. Those who know how hard the manoeuvering can be should be wary of trying to emulate a potential and achieved speed of 120 km/h. The FEI-X1, built at a Brazilian university in 1968 with a rear propeller, was a pleasant-shaped convertible, but could not hold its ground against its competitors. An American Beetle never got beyond the experimental stage: With army insignia on the doors and fat tires, it acted the tough guy when it started up its rear propeller and tuned-up engine. It probably made a lot more sense to consider a plane that had retractable wings and could also be used as a car, where earthbound propulsion took place not by means of propellers, but of powered wheels. Ideas such as a 1946 Colli flying car, a Corvair model 118 built the same year, a 1949 Molt Taylor flying car, or a recent project like the 2014 Terrafugia with pivoting propellers for a helicopter-like start are further high-flying hopes for ways of escaping a traffic jam. And even as late as 2017 the Airbus company believes that a drone-like aircraft can hang a capsule from its belly, which, mounted on a chassis, previously served as a car. All ideas of this type suffer from the strict laws that govern air traffic and will hardly produce the relief on the roads that Sunday inventors and hopeful companies come up with. In the end, we would have something out of Fritz Lang's 1926 film *Metropolis*. There, the sky was chockfull of flying machines, exaggerated as such experiments and dream visions of the future often are when put into practice.

Those who were grounded in reality were far, far more successful. The number of built cars rapidly increased in the 1920s. For instance: worldwide production of vehicles in 1920: 2 382 573, rising to 4 133 437 ten years later, and reaching 77 620 127 in 2010, i.e., in a world shaped by cars.

85. 2014 geht der Terrafugia in die Luft. (Photo: Terrafugia.)
86. Mautstation auf Chinas Autobahn. (Photo: Reuters.)
87. Ein US-Parkplatz um 1970 zeigt Farbentrends. (Photo: simoncamillieri.)
88. So enden alle – der Schrottplatz. (Photo: Bundesumweltamt.)
89. Parkplatz in Beverly Hills, um 1920. (Photo: Archiv Autor.)
90. 1955: Mercedes 220, fertig zur Auslieferung. (Photo: Daimler.)

hüten, den möglichen und auch erzielten 120 km/h nachzueifern. Der mit einem Heckpropeller ausgestattete FEI-X1 einer brasilianischen Universität aus dem Jahre 1968 hatte eine gefällige Form als Cabrio, sich aber nicht behaupten können. Aus dem Bastelstadium ist ein US-Käfer nicht hinausgekommen, der mit Wehrmachtsabzeichen an den Türen und dicken Reifen auf halbstark machte, wenn er seinen Heckpropeller mit getuntem Motor anwarf. Da waren Überlegungen, ein Flugzeug, das einklappbare Tragflächen hatte, auch als Auto einsetzen zu können, eventuell sinnvoller, wenn der erdgebundene Vortrieb nicht mit den Propellern, sondern mit angetriebenen Rädern erfolgte. Ideen wie ein Colli Aeroauto 1946, ein Corvair-Modell 118 aus demselben Jahr, ein Molt Taylor Aerocar von 1949 oder eine junges Projekt wie der Terrafugia von 2014 mit schwenkbaren Propellern für hubschrauberähnlichen Start sind weitere hochfliegende Hoffnungen, wie man einem Stau entkommen könnte. Und selbst der Airbus-Konzern glaubt noch 2017 daran, daß ein drohnenähnliches Fluggerät sich eine Kapsel an den Bauch hängen kann, die zuvor auf einem Fahrgestell als Auto diente. Alle Ideen dieser Art leiden unter den strengen Gesetzen zum Thema Flugverkehr und werden kaum jene Entlastung auf den Straßen bringen, die sich die Bastler und hoffnungsvollen Firmen einfallen lassen. Am Ende sähe es aus wie im Film *Metropolis* von Fritz Lang aus dem Jahre 1926. Da hing der Himmel voll von fliegendem Gerät, übertrieben wie so oft bei realisierten Versuchen und geträumten Zukunftsvisionen.

Wer auf dem Boden der Realität weiterfuhr, hatte mehr, viel mehr Erfolg. Die wachsende Zahl gebauter Automobile wuchs in den 1920er Jahren rapide an. Zahlenbeispiele: weltweite Fahrzeugproduktion 1920: 2 382 573, zehn Jahre später 4 133 437, 80 Jahre später, also mitten in der automobilgeprägten Welt 2010: 77 629 127. Photodokumente sind ein probates Mittel, um nachzuweisen, welche Farben, welche Formen vorherrschten, wie flächenzehrend Bewegungsflächen für Autos sind. So zeigt das Beispiel aus den USA vom Beverly Hills Speedway Parking, daß Schwarz so ziemlich die einzige, noch von Henry Ford geförderte Farbe war, auf die sich die Fließbankproduktion in den ersten Jahrzehnten festlegte, weil sie Zeitersparnis beim Lackieren und der Trocknung brachte – mit dem Nachteil, daß einzig die Kühlermaske die Marke verriet, wenn dies der schnelle Blick erlaubte.

Selbst das Bild vom Auslieferungshof bei Mercedes 1955 macht deutlich, daß die damalige Oberklasse, wir sehen hier nur 220-S-Modelle, einen Vorläufer der E-Klasse, in der Masse ihre Exklusivität verliert und sich mit Schwarz in die Seriosität rettet. Da kann der Parkplatz einer amerikanischen Großstadt 20 Jahre später beweisen, daß Buntfarben, Zweifarbenlackierung und modisch aufkommende Vinyldächer zum Trend wurden und leider eine weniger markenspezifische Identifizierung mittels Design erschwerten, weil sich nach Heckflossen und Chromgebirge in den Fronten eine ansatzweise Formenklarheit zeigte. Und daß der Wohlstand fast keine Grenzen kennen will, zeigt eine Mautstation in China, das seit 2000 mächtig aufgeholt hat und heute die Nation mit der weltweit höchsten Autoproduktion ist. Was erschaffen wird, muß eines Tages den Weg allen Irdischen gehen. Trotz Recycling, trotz gesetzlicher Entsorgungsvorgaben, trotz Export ausgedienter Kaleschen in arme Drittländer bleibt der Schrottplatz die Endstation.

Diese Station als Inspiration kann naheliegen, wenn sich Künstler damit beschäftigen und fragen: Ist die böse Seite des Autos offenzulegen, oder macht man sich lustig über jene Konstruk-

Photo documents are an effective way of verifying what colors, what forms were predominant, how much land was used up by roadways for cars. Thus, for instance, a U.S. example from the Beverly Hills Speedway parking area shows that black was pretty much the only color – promoted by Henry Ford – to which assembly line production was committed in the early decades because it meant saving time while painting and drying. The disadvantage was that only the radiator grille revealed the brand if people had time for a quick look.

Even the picture of the Mercedes delivery yard in 1955 clearly shows that the top-of-the-range class of the period – we see only 220-S models here, a precursor of the E-Class – loses its exclusivity in the mass market and, by opting for black, takes refuge in respectability. Meanwhile a parking lot in an American metropolis 20 years later demonstrates that bright colors, two-tone bodywork and stylish vinyl roofs became a trend and unfortunately made it difficult to tell apart different brands based on the design, because after tail fins and mounds of chrome in the front ends of the cars a clear form was beginning to emerge. And a toll station in China shows that there are almost no limits to prosperity: Since 2000, that country has definitely caught up and is today the nation with the highest car production in the world. Everything that is created must one day go the way of all creation. In spite of recycling, in spite of waste management laws, despite the fact that we export junk cars to poor third world countries, the scrap yard is still the end of the line.

The scrap yard is the obvious inspiration of artists who ask: Should we lay bare the negative side of cars, or should we make fun of a device that gives us mobility but can end in chaos? A

85. In 2014 the Terrafugia takes off. (Photo: Terrafugia.)
86. Toll station on China's freeway. (Photo: Reuters.)
87. An American parking lot shows color trends circa 1970. (Photo: simoncamillieri.)
88. That's where they all end up – in the junkyard. (Photo: German Federal Environmental Agency.)
89. Parking lot in Beverly Hills, c. 1920. (Photo: autor's archive.)
90. 1955: Mercedes 220, ready for delivery. (Photo: Daimler.)

tion, die uns Mobilität verschafft, aber im Chaos enden kann? Ein plattgewalzter Fiat 500 von Ron Arad oder ein sich um einen imaginären Baum wickelndes Wrack des Künstlers Dirk Skreber wären so zu verstehen. Oder Plastiken wie die in Berlin von Wolf Vostell, der seinerzeit einige Cadillacs einbetonierte. Und weltweit wurde das Bild der in die Erde gerammtem Caddies zum Wahrzeichen des Autowahns. Die sogenannte Cadillac-Ranch im Potter County, nahe der Stadt Amarillo in Texas, gibt es seit 1974. Modelle aus den Jahren 1948 bis 1963 stecken die Nasen in den Dreck, das Heck in die Luft und können gerne als Grabsteine für die Heckflossenzeit verstanden werden. Diese Idee hat etwas Humoriges, und dabei wird das Gestalten von Automobilen in Heimwerkertätigkeit schier grenzenlos. Kann der Schwanwagen von 1910 noch falsch verstanden als stolzer Beweis werden, ein teures Auto sein eigen zu nennen, so läßt der Österreicher Erwin Wurm 2015 einen kleinen LKW rückwärts die Wand hochklettern, was jeder für sich frei interpretieren darf, ebenso seine verfremdeten PKWs der schnelleren Art, die sich einem Hefeteig gleich aus ihrer Grundform verabschieden.

Da hatte sich 2007 der in England geborene Amerikaner Andy Saunders ganz andere Mühe gemacht, ein Auto zu verfremden. Einerseits war er angetan von der liebenswerten Einfachheit eines Citroën 2CV, dessen Spitzname »Ente« schon die große Zuneigung erkennen läßt. Andererseits trieb ihn die Idee um, wie ein von Picasso inspiriertes Auto aussehen könnte. 2007 hatte er dem Publikum das Ergebnis präsentiert. Eine Ente mit in allen Bereichen verschobenen Details, dabei so zurückhaltend, daß sie ihre Identität nicht verlor, aber Bilder produzierte, die nur ein Lächeln provozierten. Nichts vom Schrecken, nichts von den Gefahren des Autos. Vergleichsweise harmlos, ohne Anklage und noch humorvoll waren 2017 die Elektromobile i3 und i8 von BMW. Hier reichte es, die Außenhaut und das Interieur im Memphis-Stil zu garnieren. Sie waren nett, aber wenig bedeutungsvoll wie so viele BMWs in den letzten Jahrzehnten, die die Münchener Marke zum Bemalen bekannten Künstlern überließ, mehr PR-Spaß als ernsthafte Auseinandersetzung auf künstlerischer Ebene.

Bar jeder künstlerischen Absicht hatten Bastler und kleine Karosseriebaubetriebe versucht, sich ein eigenes Bild vom Auto zusammenzuschweißen. In Zeiten des Mangels, also nach dem Ersten, aber besonders nach dem Zweiten Weltkrieg, als der Karosseriebau schon einiges dazugelernt hatte, begaben sich viele in die Werkstatt, um etwas Sinnvolles auf drei oder vier Räder zu stellen. Was dabei betroffen macht, ist der unbändige Fleiß, mit dem sie sich einer solchen Aufgabe stellten. Inspirationen gab es zuhauf und oft in falsch verstandener Formfindung. Es war logischerweise peinlich, einen Straßenkreuzer amerikanischer Formprägung auf knapper Länge zu imitieren. Oder, wie das Beispiel des Schwimmwagens Herzog Conte von 1979 aus Hessen beweist, daß auf einem Ford-Granada-Fahrwerk nicht unbedingt eine Motoryacht zu gestalten ist, ergo: Keiner wollte diese Idee kaufen. Wie anders ist das Land der unbegrenzten Möglichkeiten! Ohne TÜV und ohne allgemeine Betriebserlaubnis in der Schärfe wie in Deutschland bleiben dem Bastler alle Möglichkeiten der Verspieltheit. Sei es, daß er die Karosserie zuhängt mit allem, was

91. 2011: Crash Car von Dirk Skreber. (Photo: amusingplanet.)
92. 2013: Pressed Flowers von Ron Arad. (Photo: Arad Ass.)
93. Der Hanomag »Kommißbrot« als Rennversion von 1928 bräuchte nach einem Unfall einen Korbflechter. (Photo: Adrian Kot.)
94. Erwin Wurms Truck geht 2015 die Wände hoch. (Photo: ZKM.)

91. 2011: Crashed Car by Dirk Skreber. (Photo: amusingplanet.)
92. 2013: Pressed Flowers by Ron Arad. (Photo: Arad Ass.)
93. After an accident, the 1928 racing version of the Hanomag »Kommissbrot« would need a basket weaver. (Photo: Adrian Kot.)
94. Erwin Wurm's 2015 truck drives up the walls. (Photo: ZKM.)

flattened Fiat 500s by Ron Arad or a wreck wrapped around an imaginary tree by the artist Dirk Skreber must probably be understood in this sense. Or sculptures such as the ones in Berlin by Wolf Vostell, who once set a few Cadillacs in concrete. And all over the world the image of the Caddys rammed into the ground became a symbol of automania. The so-called Cadillac Ranch in Potter County, near the town of Amarillo, Texas, has been in existence since 1974. Models from the years 1948 to 1963 are stuck in the dirt, nose down and rear end in the air, and may be understood as tombstones for the tailfin era. There is something humorous about this idea. At the same time, there has seemingly been no end to the number of do-it-yourselfers who have all been designing cars. There is the Swan Car of 1910, apparently the proud proof that its possessor owns an expensive car. And in 2015, the Austrian Erwin Wurm designs a little truck that climbs a wall backwards up, God only knows why, as well as a number of faster, weirdly misshapen Wurm cars that balloon like lumps of dough.

In the year 2007, British-born American Andy Saunders had entirely different ideas of how to de-familiarize a car. On the one hand, he loved the charming simplicity of a Citroën 2CV, whose very nickname – the »Duck« – expresses why he is so fond of it. On the other hand he wondered what a car inspired by Picasso might look like. In 2007 he presented the public with the result. A Duck with all its details shifted out of place, and at the same time so reserved that it did not lose its identity but produced images that merely provoked a smile. Nothing terrifying, no allusion to the risks posed by a car. Comparatively harmless, not accusing, and humorous to boot were BMW's 2017 electromobiles i3 and i8. Here it was enough to decorate the outer paneling and the interior in the style of Memphis. They were cute, but not very significant, like so many BMWs in recent decades that the Munich manufacturer commissioned well-known artists to paint, more a PR joke than serious artistic scrutiny.

Without any artistic intentions, nonprofessional tinkerers and small body manufacturers had tried to come up with their own idea of a car. In times of scarcity, i.e., after World War I, but particularly after World War II, when body manufacturing had already reached quite an advanced stage, many went to work and designed three- or four-wheel vehicles that made good sense. What is so amazing is the boundless effort they put into the task. There were tons of inspirations, and often they misunderstood the design rationale. Of course it was embarrassing to see an imitation of an American-designed road cruiser on too short a chassis. Or, as proven by the example of the 1979 amphibious vehicle Herzog Conte from Hesse, to realize that you can't necessarily put a motor yacht on a Ford-Granada chassis and suspension, which is why nobody wanted to

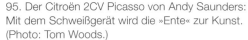

95. Der Citroën 2CV Picasso von Andy Saunders: Mit dem Schweißgerät wird die »Ente« zur Kunst. (Photo: Tom Woods.)
96. Der BMW i3 Memphis Style Edition zeigt 2017 mehr Style als Kunst, ist aber BMW-Tradition. (Photo: BMW.)
97. Wann und wo unbekannt, aber USA. Ein Mix aus den 1930er und 1950er Jahren – geht dort immer. (Photo: Archiv Autor.)
98. Dieses Spohn-Cabrio hatte 1957 noch nie einen Designer um Hilfe gefragt. (Photo: Conceptcarz.)
99. 1979 glaubte der Hersteller Herzog, sein Schiff auf Ford-Beinen werde garantiert erfolgreich. (Photo: Hersteller.)
100. Der Zubehörhandel bot für den Renault Dauphine fast zu viel Chrom. (Photo: Baboon.)

das verchromte Ersatzteillager hergibt, sei es das professionelle Zusammenlöten von Phantasievehikeln für Phantasiefilme wie die Batman-Autos, sei es die Verdoppelung von vier Rädern auf acht, um den gestreckten Cadillac auf der Straße zu halten. Amerika, du hast es besser, wenn Freaks ans Blech wollen.

Brav, wie wir sind, hatten wir schon vor Jahrzehnten ein Hanomag-Kommißbrot mit Korbgeflecht-Karosse probiert, nur um für die sportliche Version Gewicht einzusparen. Oder 1957 mit einer Spohn-Karosserie den Versuch unternommen, den Concept-Car Buick LeSabre von 1951 nachzuahmen, was völlig danebenging, sei es in der Form, dem Chromschmuck oder im Detail. Und selbst wer nicht Karosseriebauer werden wollte, konnte aus dem unerschöpflichen Angebot chromglänzenden Zubehörs sein Auto so verunstalten, daß nur geübte Augen die Marke erkannten, in diesem Fall eine Renault Dauphine. Dagegen meinte es Mercedes 1997 mit seinem F 300 (F für Forschung) ernst, denn dieses Auto mit Neigetechnik sollte fahrdynamische Tests absolvieren, auch wenn sein Auftritt nur das war: nämlich skurril.

go for this idea. How different is the land of unlimited opportunities! Without technical inspections and without the need for a general operating license as strict as the one required in Germany, Sunday tinkerers have endless possibilities to be playful. Some deck out the bodywork with every chrome spare part they can lay their hands on, others – professionals – solder together fantasy vehicles for fantasy films, like the Batmobiles, and others still double the number of wheels to eight in order to make the elongated Cadillac hold the road. In America, freakish innovators have always had so much more liberty.

Being the dutiful designers that we are, we Germans decades ago tried out a Hanomag »Kommissbrot« [square loaf of dark army bread] with a wickerwork body, merely to keep the weight down for the sports version. Or, in 1957, came up with a Spohn body in an attempt to imitate the 1951 concept car Buick LeSabre, which was an utter flop as regards form, chrome-plated trim and detail. Even people who did not want to build the body themselves, choosing from the inexhaustible supply of gleaming chrome accessories, managed to disfigure their car so badly that only experienced eyes recognized the make, a Renault Dauphine in this case. On the other hand Mercedes, in 1997, was serious about its F 300 (F for *Forschung*, research), for this car, which had tilt technology, was intended to pass extreme driving tests, even though it looked somewhat bizarre.

95. The Citroën 2CV Picasso by Andy Saunders: A welding machine turns the »Duck« into art. (Photo: Tom Woods.)

96. The 2017 BMW i3 Memphis Style Edition shows more style than art, but is a BMW tradition. (Photo: BMW.)

97. Date unknown, place – somewhere in the U.S. A mixture of the 1930s and 1950s is always popular there. (Photo: author's archive.)

98. This 1957 Spohn convertible had never asked a designer for help. (Photo: Conceptcarz.)

99. In 1979 the manufacturer, Herzog, believed his ship on Ford legs was guaranteed to be successful. (Photo: Manufacturer.)

100. The accessories trade offered had almost too many chrome parts for the Renault Dauphine. (Photo: Baboon.)

101. 1920 reduzierte sich der Slaby Behringer mit Elektroantrieb auf das Wesentliche. (Photo: Audi.)
102. 1910 kennt der AC Sociable aus England nur nützliche Pragmatik. (Photo: Simon Geoghegan.)
103. 1952 hätte man vom Mochet 125 luxe aus Frankreich schon mehr erwarten dürfen. (Photo: Classicdriver.)
104. Auch der Kleinschnittger F98 von 1952 hätte mehr bieten dürfen. (Photo: Hersteller.)
105. 1942 baute der Franzose Paul Arzens das elektrische Ei (Œuf électrique). (Photo: New York Daily News.)

5. Form und Technik machen sich klein

Das Auto zählt nach dem Eigenheim zur teuersten Anschaffung eines Haushalts. Am Anfang war die Faszination groß, der Preis hoch. Das eine bewahren, das andere verkleinern, ist seither die Herausforderung. Erste Ansätze übten sich in einfachsten Konstruktionen, leichter Bauart und unkomplizierter Technik. Das beflügelte den Drang nach individueller Mobilität, befreit vom Zwang der Schiene und vom Zwang des Fahrplans. Der automobile Weg gab sich wie eine Art Ableger des Fahrrads. Schmale Speichenräder, einfachste Hüllen in glattflächiger Ausformung, was meist den Sperrholzplatten zugute kam, oder nur stabile Stoffbekleidung boten mäßigen Wetterschutz, ein Dach war oft unüblich und zu teuer. Der englische AC Sociable von 1910 auf nur drei Rädern war eher ein fahrbares Sofa mit schräger Frontpartie als rudimentäre Karosse, die rein zufällig an den 40 Jahre späteren Unimog erinnert und kaum aerodynamischer Anpassung gehorchte. Dies ist auch vom elektrisch angetriebenen Slaby-Beringer von 1920 zu sagen. Eher Seifenkiste als kleines Auto, konnte er mit minimalen Kotflügeln Mobilität der einfachsten Art mit nur einem Sitz garantieren. Das schonte die Batterie, machte aber einsam. Und der Fahrer hatte anstelle des Lenkrads nur ein Lenkgestänge, das per Hebel die Fronträder einschlagen ließ. Einfacher wäre es wohl nur, zu Fuß zu gehen.

Auch elektrisch, aber Lenkrad und futuristische Kugelkarosse mit großen Flächen aus Acrylglas machten aus dem »Œuf« des Franzosen Paul Arzens von 1942 ein »Ei«, so die Übersetzung. Es kann als Kleinstwagen-Klassiker gelten. Die Geometrie weiß, daß die Kugel die kleinste Oberfläche mit dem größten Volumen hat; diese Erkenntnis können die ganz Kleinen nutzen, wie sie später leicht abgewandelt in der Serie zu sehen waren. Das Ei aber blieb ein Einzelfall, ebenso wie weitere Arzens-Modelle. Anders machte es die französische Kleinwagenmarke Mochet. Hier war das automobile Prinzip Vorbild für weitere Modelle: Motorhaube, Kotflügel schräg auslaufend und mit Schürzen verfeinert, schräge Frontscheibe, aber ausgeschnittene Flanke zum Einsparen der Tür und für leichteren Einstieg, so zeigte sich 1952 der Mochet CM-125 Luxe. Zeitgleich hatte der deutsche Kleinschnittger dieses Prinzip verfolgt, wobei der Namensteil »klein« nichts mit dem Winzling zu tun hatte, es war der Name des Konstrukteurs. Paul Kleinschnittger hatte seinen Prototyp schnell für die Serie optimiert und schob ihn 1950 als F 125 ins Rampenlicht, befeuert mit 125 cm, die 6 PS und eine Spitze von 70 km/h erlaubten. Das Autochen mit dem prägenden Grill bestand aus vier Einzelflächen, die man mit dem kleinen Finger deformieren konnte. Der F 125 war so leicht, daß eine kräftige, am Heck zupackende Hand das Einparken erleichterte, der Rückwärtsgang konnte dadurch entfallen.

Wie fährt man rückwärts, ohne eine Gang zu bemühen? Das war bei einigen deutschen Winzlingen mit verstellter Zündung möglich. Dieser Spaß verlangte Konzentration und fahrerisches Können, denn der rückwärts drehende Motor konnte in allen Gängen verdammt schnell rückwärts rollen, so der Messerschmitt KR 175 von 1952 und weitere Modellbrüder, die 1957 als »Tiger« auf vier Rädern manche Mittelklasse abhängten, aber auf den Namen verzichten mußten, weil ausgerechnet der LKW-Hersteller Krupp diesen bereits nutzte. Allen Messerschmitts gemeinsam war die flugzeugähnliche Acrylglaskanzel, die seitlich aufklappbar war. Das prägte den Spitznamen »Schneewittchensarg«. Wer in ersten Jahren des Wirtschaftswunders etwas besser verdiente, konnte sich seit 1955 ein richtiges Auto leisten. Das heißt, nur rein optisch. Das Goggo-

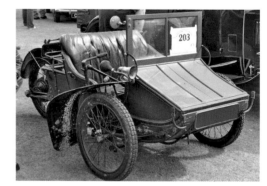

5. Form and technology go small

The biggest purchase a household will ever make is a home, and the second largest is a car. Initially the fascination was great, but the price was high. Ever since, the challenge has been to keep the fascination and reduce the price. In the early days the focus was on the simplest of designs, lightweight construction and uncomplicated technology. This inspired the urge for individual mobility, free of the constraint of having to go by railroad and keeping to its schedule. Automobiles initially looked something like spin-offs of the bicycle. Narrow spoked wheels, very simple slip-stream bodywork, which usually was of advantage to the plywood panels, or merely permanent, sturdy fabric cladding provided fair to middling protection from the weather, for a roof was often not customary and too expensive. The 1910 British AC Sociable, which only had three wheels, was more like a drivable sofa with an angled front section than a rudimentary coach, which by pure chance is reminiscent of the Unimog 40 years later and could hardly be aerodynamically adapted. This is also true of the electrically powered Slaby-Beringer (1920). A soapbox rather than a small car, with minimal fenders, it was able to provide only the simplest kind of mobility, having only one seat. This was easy on the battery, but felt lonely. And instead of a steering wheel the driver had only a steering linkage, which steered the front wheels by means of a lever. The only thing that would have been simpler would have been to walk.

The 1942 »Œuf« of the French industrial designer Paul Arzens, also electric, had a steering wheel and a futuristic spherical body with large acrylic-glass surfaces, which made it an »egg«, which is the English translation of its French name. It may be considered a classic minicar. It's a geometric fact that a sphere has the smallest surface with the largest volume; this feature can be utilized by the very smallest automobiles, of a type later seen slightly modified in series production. However, the Egg was to be an isolated case, as were later Arzens models. Mochet, the French subcompact brand, did things differently. Here the automotive principle was the model for further designs: a hood, fenders that ended at an angle and were enhanced with aprons, a slanting windshield, but with the side cut out to save having a door and to make entering easier – that was the 1952 Mochet CM-125 Luxe. Concurrently, the German Kleinschnittger followed the same principle; here the prefix »klein« (German for »small«) had nothing to do with the minicar itself – Kleinschnittger was the name of the designer. Paul Kleinschnittger had quickly streamlined his prototype for serial production and in 1950 turned the spotlight on his F 125, which was fired by a 125-cc engine that gave it 6 hp and a top speed of 70 km/h. The tiny car with the distinctive grille consisted of four individual panels that could be deformed with your little finger. The F 125 was so light that parking was made easier if a strong hand grabbed its front part, as you did not need to go into reverse.

How do you back up without shifting gears? That was possible in a few German minis when the ignition was not aligned. This trick demanded concentration and driving skill, for the motor, which was turning backward, could roll backward in all gears pretty damn quickly, as in the case of the 1952 Messerschmitt KR 175 and other similar models that in 1957, as »tigers« on four wheels, outpaced quite a few midsize cars, but were not allowed to be called that because the truck manufacturer Krupp, of all people, was already using this appellation. Common to all Messerschmitts was the airplane-like acrylic-glas dome that was laterally hinged. This gave rise to the nickname »Schneewittchensarg« (Snow White's Coffin). After 1955, people who had higher incomes during the first years of the German economic miracle could afford a real car. That is, only from a purely visual point of view. The Goggomobil, built by the agricultural machine manufacturer Glas in Dingolfing, was a classic three-box design: hood, body, a trunk that wasn't a trunk, for it housed a two-stroke engine, recognizable by tiny air slots in the rear flanks and an aluminum-framed air box integrated in the logo. Its crowning success, namely the fact that the passengers got home without being drenched, was a coupé version with a huge wrap-around front windshield, thus paying homage to the American style. A further American-style feature was found on the steering wheel, for a little lever the same size as the one for the turn signal made it possible to switch gears, say, into a simple tiptronic. People of more modest means bought the 1958 Zündapp Janus. Its front and back doors gave access to back-to-back seating; this »dos-à-dos« arrangement dated back to the early automobile era and was at the same time the origin of the car's name, for Janus, as we all know, is the double headed Roman god.

Japan, which like Germany had been devastated in the war, experienced its own economic miracle and the start of a glorious career manufacturing cars. »Start small« was the motto. No sooner said than done: for instance, with the 1960 Mitsubishi 500, a microcar whose form emulated that of midsize cars, almost like a Goggomobil. On the other hand, the so-called Eastern

101. In 1920 the electrically driven Slaby Behringer was reduced to the essentials. (Photo: Audi.)
102. The 1910 British car AC Sociable is merely useful and pragmatic. (Photo: Simon Geoghegan.)
103. In 1952 you might have expected more of the French Mochet 125 Luxe. (Photo: Classicdriver.)
104. The 1952 Kleinschnittger F98 should also have had more to offer. (Photo: Manufacturer.)
105. In 1942 the Frenchman Paul Arzens built the Electric Egg (Œuf électrique). (Photo: New York Daily News.)

mobil vom Dingolfinger Landmaschinenbauer Glas war eine klassische Three-Box-Karosse: Haube, Aufbau, ein Kofferraum, der keiner war, denn da werkelte ein Zweitakter, erkennbar an winzigen Luftschlitzchen an den Heckflanken und eine im Logo integrierte aluminiumgefaßte Lufthutze am Heck. Sein Erfolg, nämlich trockenen Hauptes heimzukommen, wurde gekrönt durch eine Coupé-Version mit riesigem, in die Seiten greifendem Heckfenster und so dem American Style huldigend. Dem wurde am Lenkrad weiter gefrönt, denn ein Hebelchen, groß wie sonst für den Blinker, machte die Gangwahl möglich, in etwa die einfache Tiptronic. Wer bescheidener war, griff 1958 zum Zündapp Janus. Hier boten eine Front- und eine Hecktür Zugang zu Sitzen, die eine Rücken-an-Rücken-Position ergaben, die Anordnung »dos-à-dos« aus ersten Automobiljahren und gleichzeitig Namensgeber für das Mobil, denn Janus ist bekanntlich der doppelköpfige Römergott.

Japan, ähnlich kriegszerstört wie Deutschland, erlebte ebenso sein Wirtschaftswunder und den Beginn einer ruhmreichen Autokarriere. Klein zu anfangen war auch angesagt; gedacht, getan unter anderem 1960 mit dem Mitsubishi 500, einem Autochen mit formaler Sehnsucht nach der Mittelklasse, fast wie ein Goggomobil. Wie schwer hatte es da der sogenannte Ostblock! Vorgegebene Planzahlen, Materialmangel, geringes Know-how sind keine Basis für Autos mit Karriere. Aber man wußte sich zu helfen, wie es die Väter des Velorex in der Tschechoslowakei seit 1950 in ständiger Verbesserung und Motorerstarkung schafften, ein Dreirad mit Stoffverkleidung für die Basis und das Dach zu bauen. Diese Stoffidee in extremer Hightech-Version zeigte uns BMW 2008 mit der eigenartigen Studie Gina. Außer Stoff war Kunststoff für Kleinserien und Kleinautos der gute Ausweg, eine Karosserie zu bilden. Wer aber 1961 auf 134 cm Länge ein Auto auf Zwergenräder stellt, kann nur ein skurriler Engländer sein, dem es egal ist, wie Schlaglöcher bewältigt werden. Dafür nennt sich dieser Peel 50 das kleinste Serienauto der Welt, selbst dann, wenn es mit hebender Hand in die Parklücke gestellt wird. Ein Händchen für Schönheit bei Kleinwagen ist eine Kunst für sich, schade daher, daß die Ford-Studie »Berliner« von 1968 nie die Serie sah. Auch nicht serienmäßig, weil speziell für die automobile Jahrhundertfeier 1986 gebaut, war das sogenannte »Centomobil«, ein gedanklicher Ableger des Arzens-Eis, ganz durchsichtig und aufregender als die entsprechende, von Daimler geförderte Feier im Fernsehen. In diesem Hause wurde schon vorher mit kleineren Modellen experimentiert bis hin zu der halb verglasten Kiste NAFA (Nahverkehrsfahrzeug) von 1981. Seine Länge von 250 cm ist exakt das Maß des ersten Smart von 1997, der nach Antichambrieren bei großen Herstellern im Hause Daimler eine Zukunft bekam. Er ist der einzige Kleinwagen, der in Sachen Sicherheit allen Anforderungen genügt, im Design aber lieber den Verspielten gab und gibt. 2014 streckte er sich um 19 cm. Wir aber vergessen die Versionen Roadster, Coupé und den Viertürer ForFour, denn es gibt nur ein Original.

106. Das tschechische Dreirad Velorex von 1962 biegt kein Blech, sondern Stoff. (Photo: A. Praefke.)
107. 1954 glaubte der Messerschmitt KR 175 an die Flieger der Marke. (Photo: Rezbach.)
108. Das Goggomobil 250 ist 1955 die Limousinenform im Kleinstformat. (Photo: L. Spurzem.)
109. 1958 hat der doppelgesichtige Zündapp Janus seinen Namen als Designprogramm. (Photo: D. Schnabel.)
110. 1961 kann keiner kleiner: der englische Peel P50. (Photo: D. Schnabel.)
111. Der Ford Berliner beweist 1968: Designer können klein. (Photo: Autor.)
112. Mitsubishi baut 1960 Bescheidenheit als Limousine. (Photo: Mitsubishi.)
113. 1998: Mercedes A-Klasse, sein Prototyp und der Smart-Vorläufer NAFA von 1981. (Photo: Daimler.)

106. The 1962 Czech three-wheeler Velorex is made not of metal, but of cloth. (Photo: A. Praefke.)
107. In 1954 the Messerschmitt KR 175 had faith in the brand's planes. (Photo: Rezbach.)
108. The 1955 Goggomobil 250 is a miniature version of the sedan shape. (Photo: L. Spurzem.)
109. In 1958 the very name of the two-faced Zündapp Janus announced its design program. (Photo: D. Schnabel.)
110. They don't come smaller than this in 1961: the British Peel P50. (Photo: D. Schnabel.)
111. The 1968 Ford Berliner proves that designers can go small scale. (Photo: author.)
112. Mitsubishi, in 1960, builds a modest sedan. (Photo: Mitsubishi.)
113. 1998: Mercedes A-Class, its prototype and the 1981 forerunner of Smart, NAFA. (Photo: Daimler.)

block was having a really tough time. Specified target figures, shortages of material, limited knowhow do not constitute a basis for cars with a future. But the East Germans found a way out, just as the fathers of the Velorex had in Czechoslovakia since the 1950s, making constant improvements and adding a stronger engine, had managed to build a three-wheeler with fabric cladding for the base and the roof. The BMW 2008 also uses fabric in an extreme high-tech version, in its unusual Gina concept car. Besides fabric, plastic was a good alternative for building a car body intended for small-scale series and compact cars. But only an eccentric Englishman who couldn't care less about negotiating potholes would place a 134-centimeter-long car on midget wheels, in 1961. This Peel 50 calls itself the smallest series-produced car in the world; what is more, it can be manually lifted into a parking space. Designing beautiful compact cars is an art in itself. That's why it is a shame that the 1968 Ford experimental model »Berliner« never saw series production. Similarly, the so-called »Centomobil« – a conceptual spin-off of Arzens' Egg – was also not massproduced, since it had been specially built for the automobile centennial of 1986; totally transparent, it was far more exciting than the Daimler-sponsored centennial celebration on TV. Even prior to this, Daimler had experimented with smaller models, up to and including the half glazed crate NAFA (*Nahverkehrsfahrzeug* or short-distance transport vehicle) of 1981. Its 250-cm length is exactly the same as that of the first Smart car (1997), which after kowtowing in vain to the big manufacturers got its chance at the Daimler company. It is the only small car that met all safety standards but whose design was, and continues to be, playful. In 2014 it became 19 centimeters longer. But we have failed to mention the roadster, coupé and four-door ForFour versions, for only one original is in existence.

6. Form und Technik sind Aufgaben für Designer

Aller Anfang ist schwer. Automobile Gestaltung war Zusammenfügen bekannter Technikdetails, um sie funktionsfähig zu schaffen. Noch fehlte die Absicht, speziell Einfluß auf die so entstandene Form zu nehmen. Beherrschte man mit der Zeit Antrieb, Fahrwerk und Passagierplazierung, so half die Kenntnis des Kutschenbaus, alles sinnvoller, handwerklich gekonnter zusammenzufügen, damit auch das Auge sich erfreuen konnte. Nicht von ungefähr etablierte sich in Norditalien die Kunst des Kutschenbaus, so daß sich in dieser Branche logischerweise Karosseriebaumeister profilieren konnten – eine Entwicklung ähnlich wie in Deutschland, England oder Frankreich, aber mit größerer Zukunft. Die Ausbildung ähnelte einem Ingenieurstudium und war zugleich handwerklich geprägt dank überlieferter Bauprinzipien und Typologien der Kutschen. Mit der Entwicklung automobiler Fertigung und der damit verbundenen Nachfrage bildete sich das klassische Erscheinungsbild des Automobils heraus, anfänglich nur in Nuancen verschieden im Markenvergleich oder versteckt im Inneren der Wagen, wenn – oft patentgeschützte – Detailentwicklungen nur dem erfindenden Kopf gehörten. Für die Geschichtsschreibung kann das Jahr 1927 festgehalten werden, als General Motors erkannte, daß automobile Formen nicht nur der Technik, der Herstellung und der Funktion zu gehorchen haben, sondern bewußt so gestaltet sein können, daß sie allein formal eine gewisse Eigenständigkeit behaupten. GM gründete die sogenannte »Art and Color Division«. Erster Chef und so bekanntester Mann der Branche wurde Harley Earl. Er setzte speziell das freie Skizzieren durch sowie das Formen mit Clay, dem Begriff des knetmasseähnlichen Materials für den Modellbau. Earl war aber auch Begründer der »planned obsolescence«, der geplanten Frühveralterung von Autoformen, unterstützt durch jährlichen Modellwechsel. Diese Haltung trug – ganz nebenbei betrachtet – dazu bei, daß weltweit photographische Straßenszenen schnell zeitlich eingeordnet werden konnten, waren doch die Karosserien eine Art Kalender. Für Earl und GM war dieses Konzept die Basis für den Titel des weltgrößten Automobilproduzenten. 1937 wurde die »Art and Colour Division« in »Styling Division« umbenannt, was den Begriff »Styling« auf Dauer in der Welt automobiler Gestaltung etablierte. GM und Earl konnten es sich wegen der großen Erfolge ihrer Marken (Cadillac, Chevrolet, Oldsmobile, Pontiac u.a.) leisten, ab 1939 jene Entwürfe vorzustellen, die bei uns als Traumwagen, allgemein aber als »Concept-Cars« zukünftige gestalterische Möglichkeiten aufzeigten, teilweise unterfüttert mit technischen Neuigkeiten.

Der erste Concept-Car war 1939 der Buick Y-Job. Mit heutigen Augen betrachtet, eher konservativ, zu der Zeit ein flaches Auto auf dem Weg, sich der Kotflügel zu entledigen, die sich mit vielen schlanken, horizontalen Chromstreifen schmückten und den Eindruck von Tempo vermitteln sollten, vergleichsweise wie bei einer Photographie die verwischte Kontur der Karosserie. GM nutzte die Präsentation seiner Concept-Cars in landesweit aufgeführten »Motorama Shows«. Ein Pulk rot-weißer, chrombehängter Trucks brachte das Ausstellungsmaterial vor Ort, wo sich mögliche Kunden ein Bild von der Zukunft machen und gleichzeitig ihre Meinung äußern konnten – quasi als hilfreiche Vorgabe für Designer und damit Vorläufer viel später aufkommender »Car Clinics«, einer üblichen Technik, lange vor Serienbeginn mögliche Käufer über zukünftige Modelle zu befragen. Earl nutzte unterdessen nach ausreichender Vorstellung des Y-Jobs dieses Auto für den eigenen Bedarf.

Dieser GM-Styling-Erfolg fand bald Nachfolger bei allen großen Herstellern, und damit stieg die Nachfrage nach speziell ausgebildeten Designern. Noch kamen die meisten Talente aus dem Karosseriebau, nur wenige von Design-Schulen. Das bot designorientierten und ausgebildeten freiberuflichen Gestaltern die Chance, im Automobilbereich tätig zu werden. Namen wie Gordon Buehrig (Vater des Cord 810 von 1936), Brooks Stevens (Vater des Studebaker GT Hawk von 1962), Virgil Exner (Vater vieler Modelle des Chrysler-Konzerns) oder Raymond Loewy (Designer mehrerer Studebaker-Modelle) prägten Autos, die meist auch zu legendären Beispielen amerikanischer Autogestaltung wurden. Die wohl bekannteste Schule für Automobildesigner wurde das 1930 gegründete Art Center College of Design in Pasadena, Kalifornien. Viele bekannte Autodesigner lehrten und lehren dort, seit 1976 im hochmodernen Glas- und Stahlbau (Architekt Craig Ellwood), es wurde weltweit zum Maßstab für Autodesignschulen, aber mit horrenden Studiengebühren. Erwähnenswert aus den USA wären noch – neben anderen – das Pratt Institute in New York und das Center of Creative Studies in Detroit, der Hauptstadt der amerikanischen Autoindustrie. In Europa sind die bekanntesten das Royal College of Art in London, das Istituto Europeo di Design in Mailand, die Fachhochschule München und die Fachhochschule Pforzheim mit ihren jeweiligen Abteilungen Transportation Design. Andere deutsche Hochschulen haben kaum die Ausbildungskompetenz wie München und Pforzheim mit ihrem inzwischen internationalen Ruf, was den Stu-

6. Form and technology are up to designers

Every beginning is hard. Automotive design involved assembling well-known technical details in order to create a functional vehicle. Initially designers did not yet specifically intend to influence the resulting form. As they gradually mastered transmission, traveling gear and the placement of passengers, their knowledge of carriage construction helped to assemble everything more logically, with more artisan skill, so that the car would be more pleasing to the eye. It was no coincidence that the art of coach-building was established in northern Italy; as a result Italian car body builders were able to rise to prominence in this field – a development similar to that in Germany, England and France, but with a more promising future. Their training resembled that of an engineer and at the same time had an artisanal aspect thanks to traditional construction principles and carriage typologies. With the development of automobile manufacture and as demand for cars rose, the classic look of the automobile emerged; initially, the various brands of cars looked only subtly different, or differences were concealed in the interior of the car when – often patented – design details were the property of the inventor. According to historians it was in 1927 that General Motors realized that automobile forms not only had to be in line with technology, manufacture and function, but could be intentionally designed so as to assert a degree of independence in a purely formal sense. GM founded the so-called »Art and Color Section«. Its first head and thus the best-known man in the industry was Harley Earl. It was he who specifically insisted on freehand sketching and on modeling with clay. But Earl was also the originator of »planned obsolescence«, encouraged by annual model change. By the way, it was as a result of this attitude that all over the world photographs of street scenes could be quickly dated, for the car bodies were a kind of calendar. For Earl and GM, this concept was the basis for the title of the world's largest automobile manufacturer. In 1937 the »Art and Color Section« was renamed »Styling Section«, which permanently established the term »styling« in the world of automotive design. Starting with 1939, because of the great success of their brands (Cadillac, Chevrolet, Oldsmobile, Pontiac et al.), GM and Earl could afford to introduce designs that we in Germany regarded as dream cars but that by and large, as »concept cars«, demonstrated future design possibilities, and sometimes featured technical novelties.

The first concept car was the 1939 Buick Y-Job. From a modern point of view, it was on the conservative side – a flat car in the process of shedding fenders that were decorated with many slim, horizontal strips of chrome to convey an impression of speed. GM exploited the presentation of its concept cars in nationwide »Motorama shows«. A bevy of red-and-white trucks, bedecked with chrome, brought the exhibits to each locality, where potential customers could see what the future held and at the same time express their opinion – as helpful guidance for designers, as it were, and thus a precursor of the »car clinics« that would be conducted much later, a common technique of surveying potential buyers with regard to future models, long before cars began to be mass-produced. Meanwhile, after presenting the Y-Job for a sufficient length of time, Earl kept this car for his private use.

This GM styling success was soon emulated by all the big manufacturers, and demand for specially trained designers grew. Most of the talented designers still came from the field of bodywork construction, while only few came from schools of design. This gave design-oriented and trained freelance designers the opportunity to work in the automobile industry. Names like Gordon Buehrig (the father of the 1936 Cord 810), Brooks Stevens (the father of the 1962 Studebaker GT Hawk), Virgil Exner (the father of many Chrysler models) or Raymond Loewy (the designer of several Studebaker models) put their stamp on cars that in most cases also became legendary examples of American car design. Probably the best-known school for automobile designers was the Art Center College of Design in Pasadena, California, founded in 1930. Many famous car designers taught and still teach there – since 1976 an ultramodern glass and steel building (architect: Craig Ellwood); it became a global standard for schools of automotive design, though the cost of tuition is horrendous. Also worth mentioning in the U.S. are – among others – the Pratt Institute in New York and the Center of Creative Studies in Detroit, the capital of the American auto industry. In Europe the best known are the Royal College of Art in London, the Istituto Europeo di Design in Milan, the Fachhochschule München (Munich University of Applied Sciences) and the Fachhochschule Pforzheim (Pforzheim University of Applied Sciences) with their respective departments of transportation design. Other German universities and colleges hardly have the same training expertise as Munich and Pforzheim with their now international reputation, a fact that benefits the students. Hardly a single student remains without a job upon graduation. That is because auto industry teachers recognize talented people early and offer them jobs.

denten zugute kommt. Kaum einer bleibt am Ende des Studiums ohne Job. Das liegt daran, daß Lehrbeauftragte aus der Automobilwelt rechtzeitig Talente erkennen und ihnen Jobs anbieten.

Deutsche Hersteller begannen erst nach dem Zweiten Weltkrieg zögerlich mit dem Aufbau einer reinen Designabteilung. Was Mercedes zunächst mit dem Begriff Stilistik bezeichnete, findet sich jetzt seit Jahren wieder in einem luftigen, großzügig ausgelegten Bau mit gefächerten Dächern (Architekten Renzo Piano Building Workshop). Bei Opel saßen die Gestalter zunächst in den alten Vorkriegsbauten und mußten für die Tageslichtkontrolle ihre Modelle auf das Dach der Werksbauten hieven. Ab 1964 durften sie das eigene weiträumige Designstudio mit ausreichend großer, vor Blicken geschützter Außenfläche übernehmen. VW-Designer haben wesentlich später die Abnabelung vom Hauptbau vollzogen. Und um ihren Designern die Atmosphäre zu versüßen, richteten sie in Potsdam in einer herrschaftlichen Villa Arbeitsplätze ein, um speziell Concept-Cars zu entwerfen. Als genereller Standard gilt heute, daß die großen Hersteller weltweit dort Studios einrichten, wo sich ihre Marke einer starken Nachfrage erfreut. So können China oder die USA neben Italien viele Beispiele nennen. Mercedes gönnt sich beispielsweise für den Bereich »Colour and Trim« – also Farbe und Ausstattung – eine feine Villa in Como, Oberitalien, nahe dem Modezentrum und Inspirationsort Mailand. Kia, Hyundai und Mazda lieben die Nähe zu Frankfurt.

Diese Beispiele hohen Aufwands für automobile Gestaltung sind Beweis und Ausdruck für deren Wichtigkeit. Die technische Zuverlässigkeit, mal das Desaster mit Dieselmotor-Manipulationen oder die unrühmlichen Verbrauchsangaben der Hersteller, gekoppelt mit gesetzlich unpräzisen Vorgaben, ausgeklammert, ist inzwischen so hoch, daß selbst früher rostbekannte Marken oder wenig zuverlässige Importe jenen Standard erreicht haben, der die Kaufentscheidung auf eine gefällige Form reduzieren kann. Das erklärt den aufwendigen Arbeitsprozeß von der ersten Papierskizze bis zur Produktionsfreigabe. Diesen Zeitabschnitt versucht man seit Jahren zu minimieren. Die Nachfrage nach ständig neuen Modellen, angefüttert durch die Fachpresse mit retuschierten Photographien kommender Modelle, nachgeholfen von den Herstellern durch Vorabberichte noch foliengetarnter Autos, zwingt zu einem rationalisierten Ablauf des Designprozesses. In frühem Stadium werden die wesentlichen Kriterien festgelegt, seien es technische Details wie Fahrwerk, Motorlage und Karosserieauslegung (Stufenheck, Schrägheck, Steilheck etc.), verbunden nicht nur mit notwendiger Familientradition der Vorgänger – siehe Beispiel Daimler mit der vertikalen Affinität und der horizontalen Homogenität –, sondern mit den Möglichkeiten neuer Motive als zukünftige Wiedererkennungsmerkmale, eine Herausforderung speziell für kommende elektrisch betriebene Autos. Haben sich in der ersten Phase mögliche Nachfolgemodelle bewerben können, werden sie schnell in Modelle im Maßstab 1:4 umgewandelt, um sie aus jeder nur möglichen Blickrichtung zu testen. Es versteht sich, daß bereits hier alle technisch orientierten Parameter wie auch gesetzliche Vorgaben und aerodynamische Grundprinzipien zu beachten sind, was Designer herausfordert, aber nicht einschränken soll, denn eine reibungslose Kooperation mit Ingenieuren aller Fachrichtungen ist die Voraussetzung für ein erfolgreiches Modell.

Die Designer haben inzwischen viele elektronische Helfer für die Umsetzung ihrer Ideen. Eine Art Muntermacher sollen sogenannte »Moodboards« sein. Hier werden Collagen mit bunten Bildchen zusammengestellt, die die Welt der möglichen Kundschaft widerspiegeln dürfen oder sollen. Übertreibungen sind dabei üblich, aber unschön, denn diese bebilderten Welten haben mehr mit Werbebildern gemein als mit der alltäglichen Realiät, falls diese nicht gerade am Pazifik, an oberitalienischen Seen oder vor Potsdamer Villen liegt. So glaubte Mercedes bei der Entschei-

114. 1948 sah es noch wie in einer Werkstatt aus, hier das englische Vauxhall-Studio. (Photo: Vauxhall.)
115. Das Werkstatt-Ambiente bei Mercedes war 1972 immer noch Usus. (Photo: Daimler.)
116. 2000 sah es im Daimler-Designstudio in Sindelfingen schon ganz anders aus. (Photo: Daimler.)
117. Zunächst weiß man nicht alles, also werden Designvarianten erarbeitet. (Photo: Daimler.)
118. Farbe kann eine Form erheblich beeinflussen, daher Proben zur passenden Auswahl. (Photo: Daimler.)

114. In 1948, workshops still looked like workshops; pictured here is the studio of the British Vauxhall. (Photo: Vauxhall.)
115. A workshop atmosphere at Mercedes was still standard practice in 1972. (Photo: Daimler.)
116. By 2000 the Sindelfingen Daimler design studio already looked quite different. (Photo: Daimler.)
117. Initially there are a lot of unknowns, so designers come up with various versions. (Photo: Daimler.)
118. Color can make a big difference to a form: Samples make the right choice easier. (Photo: Daimler.)

It was not until after World War II that German manufacturers hesitantly began setting up purely design-oriented departments. What Mercedes initially called stylistics has now for years been housed in an airy, spacious building with fan-shaped roofs (architects: Renzo Piano Building Workshop). Opel's designers were originally based in old prewar buildings and were forced to haul their models up on the factory roof for daylight control. As of 1964 they were able to move into a spacious design studio of their own with sufficiently large grounds protected from prying eyes. VW designers broke away from the main building quite a bit later. And in order to sweeten the ambience for their designers, VW set up workspaces in a stately Potsdam villa for the special purpose of designing concept cars. The general standard today is that the big manufacturers set up studios all over the world wherever their brand is in great demand. Not only Italy but China and the U.S. have many such studios. Mercedes, for instance, has a fine villa housing its »Color and Trim« department in Como, Northern Italy, near the inspiration-rich fashion center of Milan. Kia, Hyundai and Mazda love to be in the vicinity of Frankfurt.

These examples of high expenditure for auto design are the proof and expression of its importance. Technical reliability, once you factor out emission scandals or the manufacturers' shamefully inaccurate gas consumption figures, along with legally imprecise data, is meanwhile so high that even brands that were once known to rust, or less reliable imports, have reached a standard that makes decision-making easy for buyers. This is the reason for the time-consuming work of a designer, from the first paper sketch to the moment the model is approved for production. For years, there have been attempts to shorten this period. The demand for constantly new models, fueled by the trade press with retouched photographs of upcoming models, and aided by the manufacturers, who issue preliminary reports about cars that are still under wraps, makes necessary a rationalized design process. In an early stage designers establish the key criteria, for instance, technical details such as the running gear, engine position and body design (notchback, hatchback, squareback, etc.), as well as not only the necessary family tradition of previous models – for instance, Daimler's vertical affinity and horizontal homogeneity – but also possible new motifs as future brand recognition features, a particular challenge for upcoming electric cars. If potential follow-up models make the grade in the first phase, they are quickly turned into models on a scale of 1:4 so that they can be tested from every possible viewing direction. It goes without saying that even at this stage all technical parameters as well as legal requirements and basic aerodynamic principles must be taken into consideration, which challenges designers but is not intended to restrict them, for frictionless cooperation with engineers across all the disciplines is essential for producing a successful model.

Meanwhile, today's designers have many electronic aids to help implement their ideas. So-called »moodboards« are said to work as pick-me-ups. These provide collages of little colorful images that are supposed to reflect the world of potential customers. They tend to be exaggerated but unattractive, for the worlds depicted in them have more in common with advertising images than with everyday reality, unless that reality happens to be located on a Pacific beach, by North-

dung, den Smart zu bauen, daß seine Klientel die hippe Gemeinschaft der Jugend ab 20 sei. Weit gefehlt, schon Jahre nach Einführung des Winzlings lag das Durchschnittsalter der Käufer weit über Mitte 40, die wenig mit bunten Spielereien austauschbarer Kunststoff-Karosserieteile oder schrägen Innenraum-Applikationen anfangen konnten.

Der Smart kann ein Beispiel dafür sein, daß die Größe eines Autos weniger mit den Platzverhältnissen für Fahrer und Beifahrer zu tun hat. Die Vorgabe für Designer ist und bleibt der Mensch. Der ist aber so unterschiedlich gebaut, daß man ihn für das Package, also für seine Innenraumzuordnung, neben den notwendigen technischen Vorgaben wie Motorlage und Kofferraum einordnet zwischen der 5%-Frau und dem 95%-Mann. Das ist die Bandbreite zwischen der kleinsten Frau (5% aller Frauen) und dem größten Mann (95% sind kleiner). Dazu ist außerdem die ideale Sitzposition zu bedenken, bei der die Neigungswinkel von Beinen, Sitzlehne und Kopfhaltung zu berücksichtigen sind. Und die Augenellipse markiert in einem Packageplan jenen Bereich, in dem sich die Augen bewegen, unabhängig von Personengröße und Sitzverstellung, um optimale Sichtverhältnisse auf Armaturen und die Straße zu erhalten. Weitere Parameter und Vorgaben ergänzen die Aufgaben eines serienreifen Entwurfs. So erklären sich die speziell in automobilen Fachblättern gezeigten, nahe beieinanderliegenden Massen des Innenraums. Und der Laie wundert sich, wenn er im Smart dennoch ein Raumgefühl hat, als säße er in einer wesentlich höheren Klasse. Die gezeigten Maßblätter, Zeichnungen und Photographien von sogenannten »Cutaway«-Modellen, also aufgeschnittenen Modellen, als Zeichnung auch mal Röntgenbilder oder Phantombilder genannt, erklären diesen Zusammenhang.

Noch bevor ein Auto die Produktionsfreigabe erhält, werden die erfolgreichsten Vorschläge im Maßstab 1:1 gebaut, zunächst als Clay-Modell und schließlich als täuschend echtes Modell einschließlich Innenraum, zu öffnenden Türen und oft rollfähig, um in realer Umwelt allein oder im Zusammenhang mit Konkurrenzmodellen um die Gunst der Produktionsfreigabe zu buhlen. Dabei wurde schon weit vorher in ständiger Abgleichung das Außenhautmodell mit dem Innenraummodell koordiniert. Bei dieser sogenannten »Sitzkiste« werden grundsätzliche Festlegungen für die Sitzposition und Sichtbereiche getroffen. Um nicht jede Variante gleich in aufwendige Modelle umzusetzen oder um Entwürfe im fortgeschrittenen Stadium in allen Aspekten der Gestaltung zu untersuchen, helfen die »Powerwalls«, jene computergenerierten Projektionen, die mit Spezialbrillen einen räumlichen Eindruck erzeugen und so das Auto virtuell aufleben lassen. Ähnlich kann

ern Italian lakes or in front of Potsdam villas. Thus when Mercedes made the decision to build the Smart car, the company believed that its customers were a hip generation of twentysomethings. They couldn't have been more mistaken: Just years after the introduction of the tiny car, the average age of the buyers was far above their mid-40s, who scarcely knew what to do with the colorfully frivolous mix-and-match interchangeable plastic panels or the slanting interior applications.

The Smart car exemplifies the fact that the size of a car has less to do with the amount of space for the driver and front-seat passenger. Design is about people. But people are built so differently that in addition to the necessary technical specifications, such as the engine position and trunk, the space assigned to them is assumed to be adequate for people ranging between the 5-percent woman and the 95-percent man. This is a range between the smallest woman (5 percent of all women) and the tallest man (95 percent are shorter). Moreover, the ideal seated position must be taken into account, bearing in mind the angle of inclination of legs, seat backrest and head position. And in the design package the eyellipse denotes the range of eye movement, independent of a person's size and seat adjustment, in order to produce optimum visibility of instruments and of the road. Additional parameters and specifications make up the rest of the functions of a production-ready design. This is the explanation for the close-lying sections of the interior specifically shown in automotive trade magazines. Sitting in a Smart, laymen are surprised when they have a sense of being in a much higher class of car. The dimension sheets, drawings and photographs shown here of so-called »cutaway« models, drawings sometimes also called X-ray images or phantom images, explain the connection.

Even before a car is released for production, the most successful proposals are built on a scale of 1:1, initially as clay models and finally as deceptively genuine models, including the interior and doors that can be opened; often they are able to roll, vying in real life, alone or in connection with competing models, for a chance to be released for production. Much earlier in the process the model of the car's outer skin has been coordinated with the model of its interior. In this so-

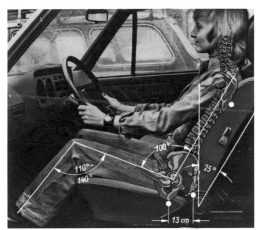

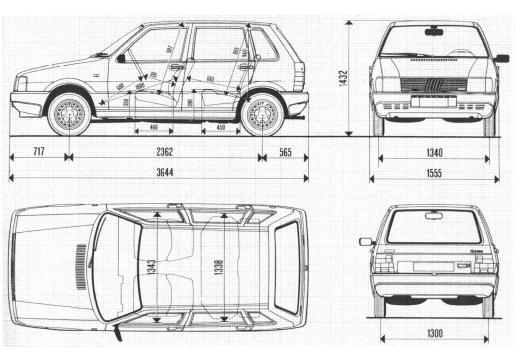

119. In 1959, on a total length of 305 cm, four passengers even have room for luggage: the Mini. (Photo: Kilbeysclassic.)
120. The 2012 U.S. Tesla S, almost 5 m long, has enough room for five. (Photo: Tesla.)
121. Almost ready to go into production: a 1:5 scale model of the NSU Ro 80. (Photo: NSU.)
122. What a designer dreamed up and what the customer might wish for: It's all on the mood board. (Photo: Jaguar.)
123. A 40-cm difference – and both must fit the driver's seat. (Photo: Volkswagen AG.)
124. The ideal seating position also has its values. (Photo: Volkswagen AG.)
125. A place has to be found for everything: The »package drawing« is helpful. (Photo: Daimler.)
126. Interior dimensions are more similar than external ones – the human body is the standard. Pictured: the Fiat Uno. (Photo: Fiat.)

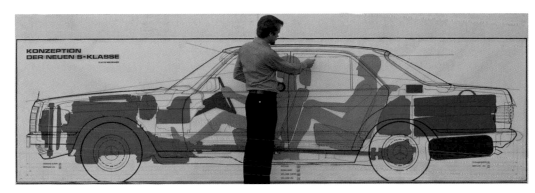

127. Opel GT-Studie in zwei Versionen, gebaut aus Clay, der »Knete« des Designers. (Photo: Autor.)
128. Bevor reale Modelle gebaut werden, kann man ihren Formenverlauf virtuell prüfen. (Photo: Daimler.)
129. Will der Designer Holz applizieren, dann muß er wissen, wo es innen verbaut werden soll. (Photo: Daimler.)
130. Eine Vielzahl der Teile hat der Designer zu verantworten. (Photo: Volkswagen AG.)
131. Man fängt klein an: Claymodell im Maßstab 1:4 oder 1:5. (Photo: Daimler.)

der Computer bereits helfen, wenn er, gefüttert mit den am Modell abgegriffenen räumlichen Koordinaten, aerodynamische Qualitäten erkundet und sein spezielles Programm darüber informiert, wie gut sich die Form unter Windeinfluß verhält. Auch kann der Computer das virtuell in Bauteile zerlegte Fahrzeug bereits jetzt in Unfälle verwickeln, um sein Crashverhalten unter unterschiedlichsten Bedingungen zu testen. Nutznießer des Computertalents sind bereits lange davor auch die Ingenieure, die für die Serienproduktion zuständig sind.

Die letzte große Aufgabe beschäftigt dann die Designer für Farb- und Materialauswahl der Ausstattung. Diese Phase nutzen manche Hersteller für eine Car Clinic, falls nicht bereits schon in einem früheren Stadium diese Befragungstechnik genutzt wurde, um sich von möglichen Kunden bescheinigen zu lassen, ob man auf dem richtigen Weg sei, allerdings unter Verschweigen der Marke oder höchstens mit Nachfragen, welcher Marke man so etwas zutrauen könnte. Diese letzte Car Clinic ist die letzte und billigste Möglichkeit, den Neuling für eine gute Zukunft zu korrigieren. Auf derartige Clinic-Möglichkeiten hatte Mercedes lange verzichtet, offenbar völlig überzeugt davon, immer das Richtige getan zu haben. Wenn beispielsweise aber entschieden wird, speziell im hohen Preissegment dem Innenraum eine Holzapplikation zu gönnen, dann belegt die dazu passende Photographie, wie umfangreich das Thema wird. Abgesehen davon, daß nicht jede als Holz identifizierbare Oberfläche auch Echtholz sein muß. Eine Geruchsprobe gäbe keine Auskunft, mögliche Echtholzvarianten sind durch mehrfache Klarlackschicht versiegelt, nur eine Feile wäre ein probates Nachweismittel, aber kein empfehlenswertes. Dafür empfehlen sich Fachleute mit feiner Nase. Ihre Aufgabe ist es, die verschiedensten Materialien und Farben so geruchsneutral oder geruchsstimulierend zu gestalten, daß mögliche Ausdünstungen davon oder von verwendeten Klebern vermieden werden. So gibt es inzwischen schon lange Sprays mit typischem Neuwagenduft für Gebrauchtwagenhändler, also die betörende Mischung von Gummi, Lack und Leder.

Sind endlich alle Hürden für den Neuling genommen, kann die Produktion beginnen. Und weil im Vergleich zu früheren Zeiten von einem Grundmodell viele Ableger zu kaufen sind – und damit sind nicht Karosserievarianten, sondern Ausstattungsvarianten gemeint –, haben die Designer auch hier vorgesorgt. Und das in einem Maße, daß kaum ein Auto aus einer Modellreihe mit einem Brudermodell identisch ist. Dies hat zur Folge, daß im Ersatzteilbereich eine hohe Zahl von Einzelelementen bevorratet werden muß. Und da kann der Hersteller eingreifen, wenn er vermeiden will, neues Blech zu biegen, aber seinem Modell ein sogenanntes »Facelift« verpassen möchte. Dank vieler Kunststoffbauteile an Bug und Heck kann das preiswert erfolgen. Sind es nur Farben, Chromapplikationen, neue Stoffe oder Lederbezüge, so wird es billiger. Der Designer ist aber auch hier ständig einbezogen, will er die Wirkung seines Gestaltungswillens bewahrt wissen.

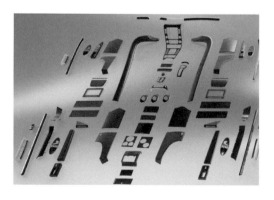

called »seat box«, fundamental decisions are made regarding the seating position and fields of vision. To avoid immediately turning each version into a costly model, or to study designs at an advanced stage in all aspects of their development, designers use »Powerwalls«, computer-generated projections that by means of special glasses create the impression of space and thus bring the car virtually to life. Similarly the computer can already help when, fed the spatial coordinates picked up from the model, it determines aerodynamic characteristics and informs its special program how well the form behaves under wind impact. Also at this stage the computer can already involve the vehicle, virtually disassembled into its component parts, in accidents in order to test its crash performance under a wide variety of conditions. Long before this, the engineers who are responsible for mass production have already benefited from using the computer's talent.

One last important task then faces the designers who choose the colors and materials used in the car's fittings. Some manufacturers use this phase for a car clinic, unless this survey technique has already been employed at an earlier stage to have potential customers confirm whether the firm is heading in the right direction, admittedly without divulging the brand name, or at most by asking which brand might come up with such a design. This last car clinic is the final and cheapest opportunity to make corrections in the new model and ensure its future success. Mercedes had long since dispensed with such clinic opportunities, obviously being totally convinced that they had always done the right thing. However, when designers make a decision, say, to add special wood paneling to the interior of an upscale car, the accompanying photograph documents how involved such a change may be – apart from the fact that not every surface that looks like wood actually has to be wood. Wood can't be identified by smell, for eventual types of genuine wood are sealed with several layers of clear varnish, where only a file would be a reliable but not recommended way of finding out. This job requires experts with a delicate sense of smell whose job it is to select a whole range of materials and paints whose smell is neutral, or to avoid certain strong-smelling materials or glues. Thus used car dealers have for a long time been using sprays with a typical new car smell – the beguiling mixture of rubber, paint and leather.

Once all the hurdles in the new model's path are gone, production can begin. And because, compared to the old days, many spin-offs of a basic model are available on the market – meaning not body types, but types of fittings –, designers have made provision for this as well. They have made sure that hardly a single car of a series of models is identical with a brother model. The result is that a large number of individual spare parts must be kept in stock by spare part dealers. And here manufacturers can intervene if they want to avoid creating new metal parts but would like to give their model a so-called facelift. Thanks to many plastic spare parts in every section of the car this can be done cheaply. If the facelift involves only paints, chrome, new fabrics or leather upholstery, it costs a lot less. Here, too, designers are constantly involved if they want to make sure the effect of their creative intention will not be diminished.

127. Opel GT study in two versions, built out of designer's modeling clay. (Photo: author.)
128. Before actual models are built, their contouring can be tested virtually. (Photo: Daimler.)
129. A designer who wants to use wood needs to know where it is to be installed inside the car. (Photo: Daimler.)
130. The designer is responsible for a multitude of parts. (Photo: Volkswagen AG.)
131. Starting small: clay model on a scale of 1:4 or 1:5. (Photo: Daimler.)

6.1. Form und Technik schaffen Bilder

Es stimmt: Ein Bild sagt mehr als tausend Worte. Das beweist sich im Alltag, wenn eine Bauanleitung für Selbstbaumöbel besser ist als eine Montagebeschreibung chinesischer Elektronikartikel, die durch den Übersetzungscomputer gejagt wurde. Oder die Erstinformation eines Verkaufsprospekts. Hatten die Altvorderen mit Stahlstichen, Patentzeichnungen oder üppig drapierten allegorischen, manchmal sogar barbusigen Figuren ihr neues Gefährt annonciert, so lagen sie mit ihrer Absicht richtig. Das Bild war immer besser, denn nicht immer war Platz vorhanden, um zusätzlich 1000 Worte unterzubringen. Ein weiterer Vorteil des Bildes: Selbst als es noch keinen Photoshop gab, als es noch keine unretuschierten, aber dramatisch in Szene gesetzten automobilen Werbebilder gab, hatte sich das Auto gern in jene Umgebung gerollt, die die gewünschte Traumwelt des Fahrers sein sollte. Es war die feine Gesellschaft, es war das Flair der großen weiten Welt, im Frack zur Gala oder im Lumberjack per Cabrio auf dem Weg in den wilden Westen. Die Zeichentechnik half mit entsprechender Darstellungstechnik nach: entweder pastellige, flächige Bilder im Stil und Trend der frühen Jahre des 20. Jahrhunderts oder kräftige Pinselführung zur Dramatisierung einer Überlandfahrt mit engen Kurven in Richtung Bergpaß.

Die neue Sachlichkeit stellte sich ein, als das Auto ab Anfang der 1930er Jahre immer mehr zum alltäglichen Straßenbild gehörte. Die Marken waren bekannt, die Gesichter ihrer Modelle ebenso. Bilder gerieten zu großen Ausschnitten der Front, zu klassischen Seitenansichten, und wenn ausreichend Fläche oder mehrere Seiten zur Verfügung standen, wurde im durchgängigen Illustrationsstil gearbeitet. Brachten sich eine oder mehrere Personen ins Bild, so war seither das freundliche Lächeln ein Muß. Und galant war der Herr immer dann, wenn die Dame zum Einsteigen gebeten wurde. Ein alter Darstellungstrick fand schon damals willige Anwendung. War das Auto klein und sollte größer erscheinen, so plazierten die Zeichner, Graphiker und Künstler die Personen hinter das Auto. Und umgekehrt davor, wenn es die Wucht des Wagens zu mildern galt. Und was draußen mit dem Auto funktionierte, ging auch im Innenraum. Winzlinge wie ein BMW Dixi, später ein Lloyd oder ein Goggomobil besannen sich auf die Familie: vorne Vater am Volant mit Mutter daneben, hinten die Kinder, und das Bild des war Viersitzers perfekt inszeniert.

Daß die politische Atmosphäre die Bilder entsprechender Zeiten beeinflußte, beweisen besonders die hier gezeigten Opelmotive. Nicht so sehr die flächige Darstellung, eine stammt vom späteren VW-Werbegraphiker Bernd Reuters, vielmehr die Wahl der Typographie erinnerte dezent an das Leben im Dritten Reich, als die Frakturschrift Usus wurde und vergessen ließ, daß Bewegungen wie das Bauhaus bereits klare, moderne und serifenlose Schriften in den Alltag trugen. Reuters ist den älteren Prospektesammlern bekannt, denn fast alle Werbemittel zeigten die komplette VW-Modellreihe, egal ob Käfer, VW Ghia oder den Transporter. Die Darstellungen waren flächig, aber mit dezenten Glanzeffekten der Lackierung, wobei Reuters das Rundliche aller VWs kräftig betonte und damit eine Ähnlichkeit zur Photographie vermied. Das mag seiner Tätigkeit als Automobildesigner geschuldet sein.

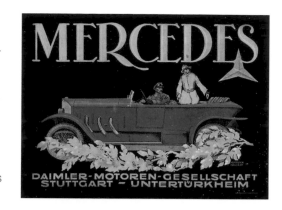

132. 1934: Deutsches Auto, deutsche Frakturschrift, beide werben für den Opel. (Photo: Opel.)
133. Mercedes ist 1920 blätterumrankt für höhere Stände mit Chauffeur. (Photo: Daimler.)
134. 1920 hilft die sachliche Zeichnung der betont gestreckten, offenen Karosserie. (Photo: Daimler.)
135. 1941 träumt die Chryslerstudie von der Zukunft, das Kriegsflugzeug kündet von der Zeit. (Photo: Chrysler.)
136. Das Armaturenbrett von 1948 zeigt: Es geht fließend in die Türseiten. (Photo: GM.)
137. Der Engländer Armstrong-Siddeley Sapphire 6 zeigt sich 1953 vornehm zurückhaltend. (Photo: Armstrong Siddeley.)
138. 1948 reicht dem Porsche aus Kärnten die Aquarelltechnik, um zu überzeugen. (Photo: Autor.)

132. 1934: A German car and German Gothic lettering advertisement for the Opel. (Photo: Opel.)
133. The 1920 Mercedes is wreathed in leaves for upper-class customers who have a chauffeur. (Photo: Daimler.)
134. 1920: a matter-of-fact drawing of the emphatically elongated, open bodywork. (Photo: Daimler.)
135. In 1941 this Chrysler study dreams of the future, while the fighter plane bears witness to the era. (Photo: Chrysler.)
136. The 1948 dashboard flows smoothly into the door sides. (Photo: GM.)
137. The 1953 British Armstrong Siddeley Sapphire 6: elegant restraint. (Photo: Armstrong Siddeley.)
138. Enough to make this 1948 Porsche from Salzburg convincing – a watercolor sketch. (Photo: author.)

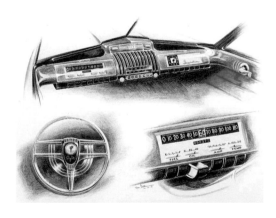

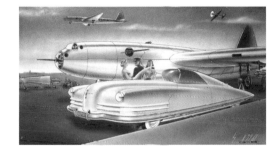

6.1. Form and technology create images

It's true: A picture says more than a thousand words. We see the proof of this every day when instructions for building do-it-yourself furniture are better than the computer-generated translation of a description of how to assemble a Chinese electronic device. Or introductory information in a sales brochure. Our forebears had the right idea when they announced their new vehicle with steel engravings, patent drawings or sumptuously draped, allegorical, sometimes even barebreasted figures. The picture was always better, for there was not always enough room to fit in a 1000 words as well. A further advantage of a picture: Even before we had photoshop, before there were unretouched but dramatically staged pictures advertising cars, automobiles were part of a driver's dream world – high society, the aura of the big wide world, a tailcoat worn to a gala, or a lumber jacket in a convertible on the way to the wild West. Drawings helped sell cars, too: either two-dimensional pastel images in the style and fashion of the early part of the 20th century or strong brushwork to dramatize a cross-country drive with tight bends as the vehicle approached a mountain pass.

The New Objectivity movement was reflected in images from a period when, beginning in the early 1930s, cars were part of the everyday street scene. The makes of the cars were familiar, and so were the faces of their models. There were now pictures of large sections of the front or classic side views of cars, and if there was enough space or several pages were available, illustrations extended over several pages. If one or several persons were portrayed in the picture, a friendly smile was now a must. And when a lady was invited to get into the car, the gentleman was always gallant. Illustrators used an old trick: If the car was small and was meant to look larger, draftsmen and graphic artists placed the people behind the car. And conversely, they were placed in front of it if the full impact of the car needed to be softened. And stratagems that worked with the outside of a car were also effective when it came to its interior. Small cars like the BMW Dixi, and later the Lloyd or the Goggomobil, used the family as an advertising ploy: Dad in front at the steering wheel with Mom beside him while the kids sat in the back, and hey presto – the four-seater was perfectly staged.

The political atmosphere of a given period influenced its images, as proven particularly by the Opel motifs shown here. Not so much the two-dimensional representation – one is by commercial artist Bernd Reuters, who later worked for VW – but rather the choice of typography was a subtle reminder of life in the Third Reich, when gothic script became common and caused people to forget that movements like the Bauhaus had already brought clear, modern and sans-serif fonts into daily use. Collectors of older brochures are familiar with Reuters, for almost all advertising showed the complete series of VW models – the Beetle, VW Ghia or the Transporter. The images were two-dimensional, but the car's paintwork gleamed subtly, and Reuters strongly emphasized the rounded features of all VWs and thus avoided pictures that looked like photographs. This may be due to the fact that he worked as an auto designer.

How different were the illustrations of designers who preferred to place their designs in futuristic settings! Their techniques were characterized not so much by naturalistic accuracy as by reduced backgrounds that focused attention on the object of desire. Depending on the illustrator's

Wie anders waren die Illustrationen von Designern, die ihre Entwürfe gerne in futuristische Umgebungen plazierten! Ihre Darstellungstechniken waren weniger durch naturalistische Genauigkeit geprägt als durch reduzierte Hintergründe, die die Aufmerksamkeit auf das Objekt der Begierde lenkten. Je nach handwerklicher Qualität der Illustratoren wirkte das Motiv entweder naiv oder überzeugend optimistisch für eine bessere Zukunft. Wie anders auch die freien Illustrationen des Automobils, wenn sie nicht dem Druck, verkaufen zu müssen, nachzugeben hatten. Aquarelltechnik oder Bleistiftzeichnung waren geeigneter, Stimmung zu generieren. Neben diesen Sichtweisen zeigt sich, daß es länderspezifische Stile gab. So waren die Engländer Meister der farbigen Darstellung. Hintergründe, Landschaften wie Schlösser oder Stadtszenen waren zurückhaltend, aber deutlich genug zu erkennen, das Motiv Auto wurde jedoch in feiner Maltechnik zelebriert, die gekonnt Lack und Chrom unterscheiden konnte, das Auto aber in fast photographischer Genauigkeit und richtiger Perspektive wiedergab. Die wohl beste Mischung aus Kommerz und Kunst beherrschte in Deutschland ein regelrechtes automobiles Malergenie. Walter Gotschke, gelernter Architekt, hatte offenbar mehr Benzin im Blut als Hausentwürfe im Kopf. Schon bald nach dem Studium entwickelte er seine autodidaktisch erworbenen Fähigkeiten, Autos in ihrem dynamischen Leben darzustellen. Die Kriegszeit hatte er in harten Szenen festgehalten, konnte aber Anfang der 1950er Jahre für Mercedes und danach für Ford Prospektillustrationen abliefern. Besonders die Modelle Ford Taunus 17m von 1957 und 1960 zeigen, daß seine Kunst perfekt war: Nicht nur das Formaltypische eines Modells war originalgetreu getroffen, auch seine Farbgebung erfaßte das Typische der markeneigenen Farbpalette. Fast konnte das Auge erkennen, ob es sich um eine Mercedes- oder eine Fordlackierung handelte, so spezifisch brach sich der Glanz in imaginärer Sonne. Seine hohe Kunst aber waren die Szenen von Autorennen, von automobilgeprägten Ereignissen und von Straßenmotiven weltbekannter Städte. Wie anders waren die Beipiele aus den USA! Hier kamen die Illustratoren für Werbematerial gern der photographischen Realität nahe. Klinisch saubere Umgebung, optimistische Gesichter, gutbürgerliche bis luxuriöse Szenen waren gefragt. Und die ohnehin schon extrem langen Schlitten wurden auf dem Papier noch weiter gestreckt. Meister auf diesem Gebiet war die Werbung der Marke Pontiac. Hier haben zwei Illustratoren als Duo gearbeitet. Der eine, Art Fitzpatrick, streckte gekonnt alle Pontiacs in die Breite und Länge. Der andere, Van Kaufmann, machte den Rest: happy people, Sonnenuntergänge, Capridörfer, Beachparties. Und selbst ein Opel Commodore durfte aus ihren Federn festgehalten werden. America at its best!

Es ist nicht übertrieben festzustellen, daß zu Gotschkes Zeit oder zu Zeiten der beiden Amerikaner nicht alle Automobildesigner dieses Talent besaßen, einem Entwurf so viel Leben einzuhauchen, daß er danach drängte, endlich gebaut zu werden. Wenige Ausnahmen sind zum Beispiel Paul Bracque (der u. a. für Peugeot und Mercedes tätig war) oder der jahrzehntelang für Opel arbeitende Hideo Kodama. Das große Talent Giovanni Michelotti hatte zum Beispiel unter anderem sehr viel zum fertigen Entwurf des BMW 1500 beigetragen, zeichnete aber eher verhalten und ungelenk. Und eine Entwurfsskizze von Raymond Loewy ist eher Arbeitsblatt als Präsentationsbild. Selbst von Nuccio Bertone, aus einer Familie von Kutschenbauern aus der Gegend um Turin stammend, wird gesagt, daß er nicht gerade ein Meister des Zeichenstifts war. Sein Talent war der Sinn für die Formen des Automobils. Und was ihn viel sympathischer machte: Bertone war einer der wenigen, der Talente erkannte, förderte und in seinem Atelier groß werden ließ. Diese Haltung in einem künstlerisch geprägten Umfeld darf eher als Seltenheit bezeichnet werden.

Dies beweist, daß ein guter Autodesigner kein guter Zeichner sein muß, zumindest nicht in jener Zeit des noch kaum gelehrten Berufsbilds. Die Zeiten haben sich seither geändert. Wer heute ernsthaft das Studium des Transportation Design (der offizielle Begriff für Autodesign) anstrebt, sollte fleißig schon vor dem Studium den Griffel so führen, daß seine Bewerbungsmappe den Nachweis liefert: Er kann nicht nur gestalten, sondern auch den Sinn und Nutzen der Idee in bildnerischer Darstellung erklären. Diese Qualitätssteigerung darf für alle weltweit tätigen Designschulen mit Schwerpunkt Transportation Design generalisiert werden. Daß diese Hürde genommen werden kann, liegt auch an den seit langem eingeführten Hilfsmitteln. So schielten wir hier vor Jahrzehnten noch nach Amerika, wo es das typische Vellumpapier gab, was für Filzstifte und Kreide ideal war, um weiche Farbübergänge zu schaffen, die einer Karosserie erst das richtige Leben einhauchten. Jeder Aquarellversuch geriet dagegen eher zu einer lieblichen Darstellung. Heute ist man viel weiter. Seitdem keine Branche mehr ohne den Computer auskommt, haben auch die Automobildesigner mit ihm eine völlig neue Möglichkeit, perfekte Renderings zu schaffen. Es gibt Zeichenprogramme, die nicht nur Grundformen in perspektivischer Richtigkeit darstellen, auch dank der räumlichen Datenstruktur, die von Modellen abgeleitet ist, sondern ebenso in der Behandlung von Farbverläufen und Glanzlichtern alle Möglichkeiten bieten, die zuvor nur auf-

139. Walter Gotschke, der beste seines Faches, zeigte 1953 den Mercedes 170s. Er zeichnete auch für Jahre Ford-Prospekte. Man konnte schon am Lack erkennen, wer der Hersteller war. (Photo: Daimler.)
140. 1968 zum Vergleich – gekonnt, aber amerikanisch übertrieben: Pontiac Bonneville Hardtop-Coupé vor beliebtem Hintergrund, wie bei allen Werbemotiven dieser Jahre der beiden Graphiker, einer zuständig für das Auto, der andere für das Ambiente. (Photo: GM.)

skill, the motif either felt naive or showed convincing optimism for a better future. Different again were the free illustrations of cars if the artists were not under pressure to sell the vehicles. Watercolors or pencil drawings were more suitable for creating a mood. Apparently styles were also country-specific. Thus the English were masters of colored illustrations. Backgrounds, landscapes such as castles or urban scenes were low-key but clearly recognizable, but the car itself was celebrated in a delicate painting technique that was skillfully able to differentiate paint and chrome, but portrayed the car with almost photographic accuracy and with the correct perspective. In Germany, what was probably the best mixture of commerce and art was mastered by a regular genius at painting cars. Walter Gotschke, an architect by profession, obviously had more gasoline in his blood than architectural designs in his head. Soon after graduating he developed his self-taught abilities – portraying cars and their dynamic lives. He had captured the war years in devastating scenes, but in the early 1950s he produced illustrations for Mercedes and later Ford brochures. The 1957 and 1960 Ford Taunus 17m models in particular show that his art was perfect: Not only did he faithfully reproduce what was typical about the form of a model, but his color scheme also caught the typical color palette of the particular make of car. The sheen of the automobile in imaginary sunlight was so distinct that one could almost recognize whether you were looking at a Mercedes or a Ford paint job. But Gotschke's greatest achievement was the scenes of auto races, of automobile-related events and of street motifs from world famous cities. How different were ads from the U.S.! Here the commercial illustrators liked artwork that came close to photographic reality. Spotlessly clean surroundings, optimistic faces, middle-class to opulent scenes were in demand. And the cars, which were extremely long as it is, were elongated even more on paper. Pontiac's advertising reigned supreme. Here two illustrators worked as a duo. One, Art Fitzpatrick, masterfully stretched all the Pontiacs in all directions. The other, Van Kaufmann, did the rest: happy people, sunsets, villages on the Isle of Capri, beach parties. Their pens even depicted an Opel Commodore. America at its best!

It s not an exaggeration to say that in Gotschke's day or in the days of the two Americans not all auto designers had enough talent to breathe so much life into a design that it practically demanded to be built. A few exceptions are people like Paul Bracq (who worked for Peugeot and Mercedes, among others) or Hideo Kodama, who for decades worked for Opel. The highly talented Giovanni Michelotti, for instance, had among other things contributed a great deal to the finished design of the BMW 1500, but his drawings tended to be cautious and clumsy. And a design sketch by Raymond Loewy is not so much a presentation image as a working image of the model. Even Nuccio Bertone, who came from a family of coachbuilders near Turin, is said to have been not exactly great at drawing. His talent was a sense for the forms of the automobile. And what made him even more endearing was the fact that Bertone was one of the few who recognized talents, encouraged them and gave them a chance to develop in his studio. This attitude in an artistic environment tends to be rare.

It all goes to show that a good auto designer doesn't have to be a good draftsman, at least not in that era, when designers had as yet scarcely learned all there was to know about their field. Times have changed since then. People who want to seriously study transportation design (the official term for auto design) today should make sure even before beginning their studies that their portfolio contains plenty of sketches to prove that they can not only design but also explain the purpose and utility of an idea in graphic representation. The latter statement generally applies to schools of design all over the world that specialize in transportation design. This hurdle can be cleared, partly thanks to aids that have long been available to designers. Thus decades ago we envied the Americans, who had a typical vellum paper that was ideal for felt-tip pens and crayons when creating soft gradations of color, which really breathed life into a car body. Attempts to do this with watercolors tended to produce merely pretty pictures. Today we've made much more progress. Since no industry can get along without computers anymore, auto designers, too, now have a completely new way of creating perfect renderings. There is drawing software that not only represents perspectively accurate basic forms, in part also thanks to a spatial database derived from models, but also offers every possibility in the treatment of color gradations and highlights which could formerly only be produced manually with the help of a spray gun, swabs of cotton wool and scads of subtly differentiated markers. Moreover, computers also make it possible to create an unlimited number of data-generated images. These are no longer stills. They are complete films that can be created with the help of the computer, breathtaking and brutal in part, as for instance in action movies like *The Fast and the Furious*. But computers can also show us the picture of a new model that we will drive in years to come now that we've been seduced into buying it.

139. Walter Gotschke, the best in his field, portrayed the 1953 Mercedes 170s. For years he also drew Ford brochures. You could tell just from the paintwork who the manufacturer was. (Photo: Daimler.)
140. 1968 by way of comparison – skillful but exaggerated, as is typical in the U.S.: a Pontiac Bonneville Hardtop Coupé against a popular background; as in all advertisements by the two artists in those years, one of them was responsible for the car, the other for its surroundings. (Photo: GM.)

141. Hideo Kodama zeigt 1985 eine Opelstudie. Er ist als Illustrator so gut wie als Designer. (Photo: Opel.)
142. 1964 entwirft Michelotti besser, als er zeichnet, hier der Triumph GT6. (Photo: Michelotti.)
143. Der Designer der Cola-Flasche, Raymond Loewy, skizziert 1961 eine Coupéstudie. (Photo: Loewy.)
144. 1955: Weißwandreifen, Flossen, Chrom, Villa, auch die Dame paßt zum Chrysler. (Photo: Chrysler.)
145. Saab zeigte 2001 den Innenraum der Studie 9.3-x in seltener Bescheidenheit der Graphik. (Photo: Saab.)

wendig in Handarbeit mit Spritzpistole, Wattebäuschen, Wischern und Unmengen an farbnuancierten Markern umsetzbar waren. Und ein übriges leistet der Computer, wenn wir an die grenzenlose Machbarkeit datengenerierter Bilder denken. Da sind es nicht mehr Stills, also unbewegte Bilder. Es sind komplette Filme, die wir rechnergesteuert gestalten können, atemberaubend und streckenweise brutal, wenn es um Actionstreifen geht wie *The Fast and the Furious*. Aber auch bei Designaufgaben wird uns das Bild eines neuen Modells vermittelt, das wir, heute zum Kauf angefüttert, in den kommenden Jahren fahren dürfen.

Was in den Designschulen häufig auffällt, sind die Aufgabeninhalte. Fast ständig werden nur extreme Fahrzeugvorgaben gemacht, die als Basis den ultimativen Sportwagen kennen. Selten ist eine ganz normale Limousine oder gar ein Nutzfahrzeug das Thema. Mag sein, daß der Sexappeal eines Sportwagens die Kreativität eher beflügelt als die Suche nach einer adäquaten Form für einen Zehntonner. Zu dieser fast einseitigen Auslegung eines Fahrzeugentwurfs kommt, daß in Sachen technischer Machbarkeit ein großer Spielraum eingeräumt wird. Mag sein, daß so etwas eines Tages baubar wird. Möglicherweise sind es sinnvolle Lösungen, die die automobile Zukunft erwartet. Selbst wenn bereits das führerlose, sich in kontrollierter Eigenregie bewegende Auto mit elektrischem Antrieb grundsätzlich neue Kriterien für den automobilen Entwurf verlangt, so sind diese Lerninhalte neben den notwendigen Fächern eher Sammelbecken für den Nachweis talentierter Ideenjäger, die die Autoindustrie vorrangig sucht, um ein formal zwar nicht ausgereiztes, aber immer schwierigeres Aufgabegebiet zu meistern. Zeichnungen, besser gesagt die Renderings, können hier Beleg dafür sein, daß – selbst als Entschuldigung geltend, es sei doch nur eine Idee auf Papier – eine Idee auf Rädern letzten Endes auch umsetzbar sein muß, um auf unseren Straßen zurechtzukommen. Soll heißen, wo begeisternde Renderings Karossen zeigen, deren Räder, niederquerschnittbereift, in Radhäusern stecken, die keinen Federweg zulassen, da kann Begeisterung schnell in Skepsis umschlagen. Und nicht selten haben dann Realität gewordene Automobile, die auf dergleichen Gestaltungsideen basieren und den Autonarr als Concept-Car beglücken, nicht nur mit Radhäusern, Radvollverkleidungen oder Lenkeinschlägen zu kämpfen, sondern negieren zuweilen auch den brutalen Alltag gesetzlicher Vorgaben, wenn ein Rücklicht nicht die notwendige Lichtaustrittsgröße hat oder die Position von anderen vorgeschriebenen Bauteilen neu, aber regelwidrig interpretiert wird. Wenn aber etwas von der ersten Skizze über Rendering und interne Modellauswahl den Weg unbeschadet in die Serie schafft und seinem Concept-Car-Vorläufer immer noch ähnelt, macht es die Designer glücklich.

The striking thing at schools of design is the projects they are involved in. Almost without exception they work only on extreme vehicle specifications based on the ultimate sports car. Rarely does the design involve a normal sedan or even a utility vehicle. Maybe the sex appeal of a sports car inspires creativity more than the search for a suitable form for a ten-ton truck. While this is an almost one-sided interpretation of what a car design is about, designers are also given tremendous leeway as regards technical feasibility. Perhaps the designs they come up with can actually be built one day. Possibly they represent viable future solutions. Even though radically new auto design criteria are already required by the driverless electric car, such projects, compared to the required courses, tend to be there for the purpose of supplying talented brains primarily sought by the auto industry so that they will master a field that, while it has not been technically exhausted, is becoming more and more complex. Drawings, or rather renderings, may here serve to document the fact that ultimately it must be possible to implement an idea on wheels in such a way that it can be driven on our roads. That is, where enthralling renderings show car bodies whose wheels have low profile tires and are in wheel wells that have zero value for suspension travel, enthusiasm can quickly turn into skepticism. And not infrequently a car based on such a design, once built and in the possession of a delighted car buff, not only has problems with its wheel wells, wheel fairings and wheel angles, but also sometimes contravenes the brutal reality of legal requirements when the illuminated area of a rear light is not large enough or the position of other prescribed component parts is reinterpreted contrary to regulations. However, if some part of the first sketch manages to survive the rendering and in-house model selection to make it into the production stage and still resembles its concept car forerunner, it makes the designers happy.

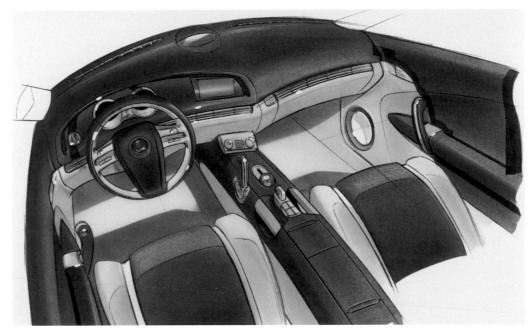

141. A 1985 Opel design study by Hideo Kodama. His illustrations are as good as his designs. (Photo: Opel.)
142. In 1964, Michelotti is better at designing than at drawing; pictured here is the Triumph GT6. (Photo: Michelotti.)
143. The designer of the Cola bottle, Raymond Loewy, sketches a study of a coupé in 1961. (Photo: Loewy.)
144. 1955: whitewall tires, tailfins, chrome, a villa – even the lady matches the Chrysler. (Photo: Chrysler.)
145. In 2001, Saab showed the interior of the design study 9.3-x in a rarely modest graphic. (Photo: Saab.)

6.2. Form und Technik bilden ein Konzept

Designer sehen ihr Glück darin, ein Auto zu gestalten, das nicht den Zwängen des Alltags unterworfen ist, sondern wagemutig den Blick in die Zukunft bietet. Autokonzerne erlauben sich schon lange Designstudios mit dem Arbeitstitel »Advanced Studio« nur für Concept-Cars, also Zukunftsstudien. Sie nehmen mittlerweile auch Fachleute auf aus ganz anderen Forschungsbereichen zum Thema Zukunft oder was dafür auf Grundlage heutiger Situation morgen möglich sein kann: technisch machbare Konzepte wie gesellschaftliche Veränderungen. Der Bereich Technik hat seinen Schwerpunkt in neuen Antriebskonzepten auf Basis der Elektrizität. Dabei ist der Sieger noch auszurufen: das batteriegetriebene Auto oder die Brennstoffzelle. Hier darf an die Naivität erinnert werden, mit der in den 1950er Jahren noch das Heil der Atomenergie gepredigt wurde. Man schreckte nicht davor zurück, sich ein reaktorbestücktes Automobil vorzustellen. Das war abwegig, ebenso wie Jahrzehnte zuvor die Idee des Propellerantriebs oder in denselben 1950er Jahren der Versuch mit Turbinen. Die heutigen Hybridautomobile werden nur eine Zwischenlösung zu anderen Antriebstechniken sein. Der reine Elektroantrieb wird sich differenzierter zeigen, weil das Brennstoffzellenkonzept größere Lufteinlässe am Bug benötigt als das Batterieauto. Viel interessanter wären neue Verhaltensmuster der Kunden. Schon heute will man erkannt haben, daß die junge Generation das Auto nicht als erweitertes Ego begreift und wenig Wert auf Prestige legt – im Gegensatz zu den PS-starken Boliden, die Auge und Ohr malträtieren. Wer weniger den Egoverstärker wünscht, könnte sich auch für neue Mobilitätskonzepte interessieren sowie für den aggressiven amerikanische Uber-Konzern, der das herkömmliche Taxikonzept durchkreuzt. Oder das Interesse verlagert sich auf Konzepte allseits erreichbarer Mietwagen, auffindbar per Smartphone. Kommt das selbstfahrende Auto in Perfektion, dann vielleicht auf Anforderung sogar vor die Tür gerollt.

Diese mögliche Zukunft wird nur schrittweise erfolgen, allein die Infrastruktur müßte große Veränderungen erfahren, bis neben einem dichten Zapfstellennetz (solange elektrisch gefahrene Reichweiten nicht dem Benziner gleichen) das Straßennetz in seiner Qualität speziell in ländlichen Bereichen dem selbstfahrenden Auto ausreichend Bewegungsmöglichkeit bietet. Auf diese Zukunft bereiten sich die Designer vor. Noch prägen ihre Vorschläge ein Autokonzept, das im Prinzip die Fortschreibung heutiger Modelle ist. Wenige Ausnahmen von Nischenanbietern versuchen sich in Entwürfen für Fahrzeuge, die nicht den individuellen Vorstellungen vom Besitz eines Prestigeprodukts anhängen, sondern bemüht sind, das ursprüngliche Prinzip des Individualverkehrsmittels umzupolen auf Formen öffentlicher Nahverkehrssysteme, reduziert auf PKW-Größe. Der Blick zurück zu den Anfängen ist dagegen ein Sammelsurium von Konzeptautos. Denn die automobilen Anfänge waren wegen der geringen Zahl einzelner Modelle nichts weiter als eben Konzepte. So ist das erste patentrechtlich zum Auto geadelte Benz-Modell von 1886 auch ein Concept-Car, wenn das Kriterium gilt, daß nur ein einzelnes Exemplar für morgen gezeigt wird. Diese Liste fortzuführen, fragmentiert die Geschichte des Automobils. Hier konzentriert man sich auf jene Konzepte, die angeblich frühzeitig Ansätze für mögliche verbesserte Automobile erkannten. Dabei wird sich Abwegiges und Sinnvolles immer wieder um Aufmerksamkeit bemühen.

1928 gab es in Frankreich ein Beispiel, das das Auto als ein praktisches, kompaktes Gefährt verstand und sich entsprechend in seinen Bauteilen reduzierte, ohne auf notwendige Karosseriequalitäten wie ein festes Dach zu verzichten. Die Voraussetzung war nicht, daß nur ein Architekt imstande wäre, dies zu schaffen, auch wenn der damals zu den bekanntesten seines Berufsstands zählte: Charles-Éduard Jeannaret-Gris, weltbekannt als Le Corbusier. Er konnte allein 17 seiner Bauten als Weltkulturerbe nennen, so die Villa Savoye von 1928, die Kapelle von Ronchamp von 1955 oder das Wohnhochhaus Unité d'habitation in Berlin von 1958. Die ersten Skizzen für das »voiture minimum« von 1928 suchten einfache Details, die Herstellung und Nutzwert optimieren sollten. Sein ausgearbeitetes Konzept wurde 1936 im Rahmen eines Wettbewerbs der Öffentlichkeit vorgestellt. Viele sehen darin den Vorläufer und das Vorbild für beispielsweise den Urkäfer, die »Ente« von Citroën oder die Karosserieformen mit aerodynamischen Prinzipien. Erst zum 100jährigen Geburtstag von Le Corbusier wurde ein Holzmodell im Maßstab 1:1 für eine Sonderschau seiner Werke im Centre Pompidou in Paris gebaut. Nachbearbeitete Zeichnungen haben aufgezeigt, wie konsequent das Minimum-Auto geometrisch aufgebaut war, mit Kreisen, Achsbezügen und Proportionssystemen. Diese Art des Gestaltens entspricht eher dem Kopf eines Architekten als der Denke eines Automobildesigners, der mehr den Zusammenklang des Autovolumens verfolgt.

Ein weiterer Architekt, Richard Buckminster Fuller, bekannt durch Leichtbausysteme und geodätische Kuppeln für großflächige Überdachungen, war bestrebt, generell eine synergetische An-

146. Der Name »voiture minimum« war 1928 Programm für den weltbekannten Architekten Le Corbusier. Während einige Kollegen beim Ausflug in das Automobildesign mehr formales Interesse am Fahrzeug hatten, durchdachte L.C. das Auto von Grund auf. Das Ergebnis war ein Zweisitzer für mobile Basisbedürfnisse, ergänzt durch überlegte Proportionen. (Photo: Design Museum London.)

146. The term »voiture minimum«, in 1928, expressed the program of the world-famous architect Le Corbusier. While some of his colleagues who ventured into car design had a more formal interest in the vehicle, L.C. conceptualized the car from the ground up. The result was a two-seater that met basic needs, plus well-thought-out proportions. (Photo: Design Museum London.)

6.2. Form and technology constitute a concept

The happiness of a designer lies in designing a car that is not subject to everyday constraints, but boldly provides a vision of the future. For many years now, car manufacturers have had design studios called »advanced studios« that are there solely for concept cars, i.e., for future studies. In the meantime they also accept future study experts from completely different fields of research, or experts on things that may be possible tomorrow based on the situation today: both technically feasible concepts and social changes. The main focus of the field of technology is on new propulsion concepts based on electricity. The winner has not been determined yet: the battery-driven car or the fuel cell. I'd like to remind readers about the naiveté with which the great benefits of atomic energy were still being preached in the 1950s. People had no qualms about imagining a reactor-fueled car. That was far-fetched, as had been the idea, decades earlier, of propeller-driven vehicles or, also in the 1950s, the experiment with turbines. Today's hybrid cars are probably only an interim solution on the way to other driving techniques. The all-electric drive will be more sophisticated because the fuel-cell concept requires larger air inlets in the front of the car than the battery car. Much more interesting would be new customer behavior patterns. Even today there are those who claim that the younger generation does not see the car as an extension of their ego and attaches little importance to prestige – in contrast to the powerful high-speed cars that torture our eyes and ears. People who have less interest in the ego boost of owning a car might also be interested in new ways of getting about, such as the aggressive American company Uber, which thwarts the traditional concept of taxi services. Or people's interest shifts to the idea of easily accessible rental cars that can be located by smartphone. Once the self-driving car is perfected, such cars might even drive up to the customer's door upon request.

This potential future will come gradually, though the infrastructure would first have to undergo great changes: In addition to a dense network of charging stations (as long as the range of electric cars does not equal that of gas vehicles), the quality of the road system, especially in rural areas, must be high enough for self-driving cars to be able to move about easily. Designers are getting ready for this future. At the moment they are still proposing a car concept that is basically an updated version of today's models. A few exceptions – niche providers – are trying their hand at designing vehicles that do not cater to the desire to own a prestige product, but rather strive to reverse the trend of individual car ownership to replace it with types of public local transport systems, reduced to passenger car size. A look back at the beginnings, on the other hand, shows us a hodgepodge of concept cars. For the early cars, because of the small number of individual models, were really simply concept cars. That was true of the first Benz model of 1886 that was patented as a motorcar, the criterion being that only a single example of a future design was shown. To continue this list would be to fragment the history of the car. Here we'd like to concentrate on those concepts that, early on, ostensibly represented the initial stages of ways that cars could be improved. This is why the present book repeatedly focuses on features that are both far-fetched and practical.

In 1928, in France, one design interpreted the car as a practical, compact vehicle and accordingly minimized the number of its components without dispensing with such essential requirements as a solid roof. Its creator had not necessarily to be an architect, though at the time he was the best known in his profession: Charles-Éduard Jeannaret-Gris, now known worldwide as Le Corbusier. No less than 17 of his buildings are included in the World Heritage list, for instance the Villa Savoye, 1928, the Ronchamp chapel, 1955, or the apartment building Unité d'habitation in Berlin, 1958. The first sketches for the »voiture minimum« in 1928 tried to keep the details simple, thus optimizing the car's manufacture and utility. His finished concept was introduced to the public in 1936 as part of a competition. Many regard this car as the precursor and model for such subsequent designs as the original Beetle, the Citroën »Duck« or the car body types based on aerodynamic principles. It was not until the centenary of Le Corbusier's birth that a wooden model was built on a scale of 1:1 for a special exhibit of his works at the Centre Pompidou in Paris. Reworked drawings show how logically the Minimum Car was geometrically constructed with circles, axial relationships and systems of proportion. This kind of design is more typical of architects than of automobile designers: The latter tend to concentrate on the harmony of the car volume.

Another architect, Richard Buckminster Fuller, known for his lightweight systems and geodetic domes, strove in general to use buildings and equipment synergistically. Cars were part of his planning concept. The 1933 Chicago World's Fair was the right place to introduce his three-wheeled oddity. The car had room for eleven persons, possibly thanks to its basic zeppelin-like

wendung von Bauten und Geräten zu schaffen. Dabei war für ihn das Thema Automobil ein Teil seines Planungskonzepts. Die Weltausstellung 1933 in Chicago war der Ort, um seinen dreirädrigen Sonderling vorzuführen. Das Auto bot Platz für elf Personen, möglich dank zeppelinähnlicher Grundform und damit vorteilhafter Aerodynamik, erreichte einen geringen Verbrauch und angeblich 193 km/h. Dank des gelenkten Hinterrads konnte das Auto auf der Stelle wenden. Seine Vorteile zerflogen bei einem Unfall auf dem Weltausstellungsgelände, als der Fahrer starb. Jahrzehnte später hatte ein anderer weltbekannter Architekt, Norman Foster, 2010 den Nachbau beauftragt. Von heute aus betrachtet, darf festgehalten werden, daß trotz hohem Tempo, geringem Verbrauch, dem Dreiradprinzip und der Baulänge keine Chance für die Serienfertigung eines Autos bestanden hätte, das einem seiner Flügel beraubten Flugzeug ähnelt.

Der erste Concept-Car mit der einzigen Absicht, neue Wege im Design zu gehen, aber bewährter Technik zu folgen, hatte 1939 Harley Earl als Chefdesigner bei GM im Modell Y-Job umgesetzt. Daß lediglich die Blechhaut progressiv zu erscheinen versuchte, sollten die mit Chromstreifen geschmückten Kotflügelschürzen belegen. Verräterisch behielt die Motorhaube die alte Höhe bei. Noch fehlte das technische Innenleben niedrig bauender Antriebe. Hier ging Chrysler 1941 mit dem Thunderbolt einen Schritt weiter. Dieser Entwurf verzichtete wie der Y-Job auf ein fest montiertes Dach, es gab lediglich ein demontables Hardtop. Schade, gerne wüßte man, was den Designern für das sogenannte »Greenhouse« eingefallen wäre, wie also Glas- und Blechflächen zur Gesamtform gepaßt hätten. So kann dem Thunderbolt attestiert werden, daß er eine konsequente Pontonform zeigte, die sich im Schwellerbereich kupferfarbenen, geriffelten Metallschmuck gönnte. Ein typischer Grill fehlte, als Concept-Car mußte er ja anders als die Serie sein. Es fällt auch nicht auf, daß die Fronthaube immer noch für den hoch bauenden Motor den Buckel machte. Thunderbolt heißt auf deutsch Donnerkeil, daher mußte die Tür den ebenfalls kupferfarbenen Donnerkeil akzeptieren.

Die Nachkriegszeit genoß den Frieden und die Freiheit, wie konnte die Autowelt das besser demonstrieren als mit neuen Karosseriestudios. In Italien begann die Karriere vieler von ihnen; eigene Entwürfe und Auftragsarbeiten brachten Geld. Frische Ideen oder inspirierende Motive aus den USA wurden sofort in Blech umgesetzt, so zum Beispiel der kleine Fiat 500 von Francis Lombardi 1950: Eine recht klassische Grundform gönnte sich ein in die Seiten greifendes Heckfenster, wie beim Studebaker Champion Starlight Coupé aus dem gleichen Jahr. Ein Jahr später überraschte der Buick LeSabre die Autowelt. Wo ein 49er Serien-Cadillac stummelähnliche Heckflossenknospen zeigte, ließ der LeSabre sie wachsen. Seine Front vergaß klassische, flächige Chromfronten und baute wegklappbare Doppelscheinwerfer, eingerahmt im zentralen Oval. Wohl zum ersten Mal war die Frontscheibe als Panoramascheibe mit negativ gestellter A-Säule zu sehen, ein weltweit nachgemachtes Motiv, das sich in der Klasse kleiner Hubräume wiederfand, so bei uns im Glas Isar 600 von 1958. Ultimatives Wiedererkennungsmerkmal der Seitensicht war die extrem großflächige, geriffelte Chromapplikation, eingerahmt von einem parabelförmigen Chromband. Zurückhaltender war Porsche 1952 mit einem Entwicklungsauftrag für eine Studebaker-Limousine. Eine klassisch ruhige Stufenheckform beweist: Je weniger Effekt, desto langlebiger die gewünschte Akzeptanz. Der Studebaker erlebte leider keine Serienproduktion.

Die war von Anfang an nicht zu erwarten, als Bertone auf dem Fahrgestell des Alfa Romeo 1900 Super Sprint 1955 den B.A.T. 5, 1956 den B.A.T. 7 und 1957 den B.A.T. 9 in automobilen Kunstformen zeigte. Sie gingen zwar an die Grenze des Ertragbaren, aber die ihnen eigene Eleganz vermied die Üppigkeit amerikanischer Beispiele und bot beste aerodynamische Qualitäten, was von den runden, stark verglasten Formen auch zu erwarten war. Das übernehmen heute Kunststoff-Bauteile, wenn komplizierte Formen gewollt sind. Mit Bertone feierte die italienische

147. 1938 kommt der Buick Y-Job noch vergleichsweise zurückhaltend auf die Straße. Auch ist die Pontonform nur zu erahnen, zu sehr sind vordere und hintere Kotflügel Solisten. Dafür zieht im Detail die Chromlinie auf der Haube ohne Unterbrechung in den zentralen Steg der Frontscheibe. Aber neu ist die Üppigkeit der dünnen Chromstege als Symbol für Tempo. (Photo: GM.)

148. Der Chrysler Thunderbolt kann 1941 alles unterm Blech verbergen. Das spart Designaufgaben für die Felgengestaltung. Und die gewölbte Motorhaube verrät den noch hoch bauenden Motor. Erst sechs Jahrzehnte später wird das Chrysler-Kupfer bei Peugeot als Chromersatz wiederentdeckt. (Photo: RM Auctions.)

149. Der Buick LeSabre macht 1951 die Panoramascheibe und die Heckflosse salonfähig, die verchromte Flanke schafft es in dieser Üppigkeit nicht in die Serie, dafür ist das Armaturenbrett eher brauchbar. (Photo: GM.)

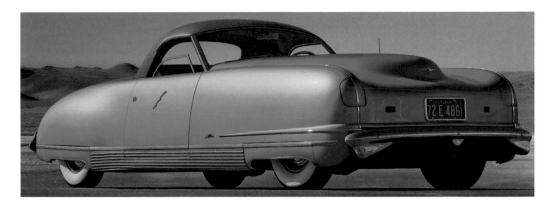

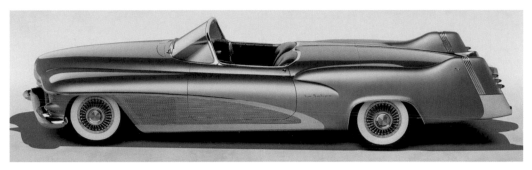

form and thus advantageous aerodynamics, had low gasoline consumption and was said to achieve speeds of 103 km/h. Due to a steered rear wheel the car was able to turn at a spot. Its advantages disappeared in a flash when the driver was killed in an accident on the grounds of the World's Fair. Many years later, in 2010, another world famous architect, Norman Foster, commissioned the building of a replica. Looking back, we can say that in spite of its high speed, low energy consumption, the three-wheel principle and its overall length there would have been no chance of manufacturing a car that looks like a plane robbed of its wings.

The first concept car whose sole intention was to break new ground in design while sticking with tried and true technology was designed in 1939 by Harley Earl, the GM's designer in chief. This was the Y-Job. The aprons of the fenders, which were decorated with strips of chrome, indicated an attempt to make the sheet-metal skin look progressive. The hood still had its original height. Technically speaking, the car lacked low-profile drives. That's where the 1941 Chrysler Thunderbolt went a step further. Like the Y-Job, this design did not have a permanently mounted roof: there was only a demountable hardtop. It's a shame – we wonder what ideas the designers would have come up with for the so-called »greenhouse« car, in other words, how glass and metal surfaces would have complemented its overall form. Thus we can attest that the Thunderbolt had a consistent pontoon shape bedecked with copper-colored, grooved metal ornamentation in the sill area of the car. It did not have a typical grille: Being a concept car it needed to be different from the production run. Nor is it noticeable that the front opening hood still humps its back to accommodate the high-profile engine. Since the car is called a Thunderbolt, the door must necessarily also feature the copper-colored lightning bolt.

As the postwar period savored peace and freedom, the car industry celebrated that fact with new automotive design studios. Many designers began their careers in Italy; their own designs and commission work brought in money. Fresh ideas or inspiring motifs from the U.S. were immediately implemented, for instance Francis Lombardi's little 1950 Fiat 500: A really classic basic shape was accompanied by a wrap-around rear window similar to that of the Studebaker Champion Starlight coupé built the same year. One year later the Buick LeSabre startled the world of car enthusiasts. While a 1949 production Cadillac had budding little stubby tailfins, the LeSabre's tailfins were far more pronounced. Its front did away with classic, two-dimensional chrome panels and featured retractable double headlights framed within a central oval. Probably for the first time, the front windshield took the form of a wrap-around window with a negatively angled A pillar, a motif that was imitated all over the world, being found in the class of cars with a smaller cylinder capacity, e.g. the German Glas Isar 600 of 1958. The ultimate distinctive feature of its side view was the extremely large, grooved chrome application, framed by a parabola-shaped chrome strip. The 1952 Porsche was more reticent, with a development contract for a Studebaker sedan.

147. The 1938 Buick Y-Job is still comparatively conservative. The pontoon form is still a mere suggestion, as the front and back mudguards try to steal the show. A chrome line continues without interruption along the hood to the central bar of the windshield. A new feature are the opulent thin chrome bars – a symbol for speed. (Photo: GM.)
148. The 1941 Chrysler Thunderbolt conceals everything under its metal bodywork, eliminating the need for wheel rim design tasks. And the curved engine hood betrays the high-profile engine. It is only 60 years later that Peugeot rediscovers Chrysler's copper as a substitute for chrome. (Photo: RM Auctions.)
149. The 1951 Buick LeSabre makes the wrap-around windshield and tailfins presentable; the chrome flank of this opulent version doesn't make it into series production. On the other hand, the dashboard definitely comes in handy. (Photo: GM.)

150. 1953 bis 1955 zeigt Bertone mit den BAT 5, 7 und 9 die hohe Kunst der Blechverformung. (Photo: Bertone.)
151. 1967 veröffentlicht VW mögliche Käfernachfolger, hier vier von 30. (Photo: Volkswagen AG.)
152. 1954 baut Ford die Rakete FX Atmos auf vier Rädern. (Photo: Ford.)
153. 1958 baut GM den Jet auf vier Rädern: den Firebird III. (Photo: GM.)
154. 1965 kann die Sachlichkeit eines zukünftigen Vans den Autonova-Fam prägen. (Photo: Caspers.)
155. Der Borgward Traumwagen fährt 1956 völlig aus der Gegenwart: zu übermütig. (Photo: Borgward.)
156. 1957 beweist Raymond Loewy: Abkehr vom Üblichen macht den BMW 507 elegant. (Photo: Conceptcarz.)

Kunst des Blechverformens ihre Höhepunkte. B.A.T. steht übrigens für Berlinetta Aerodinamica Tecnica; der Volksmund gab sich auch mit dem Begriff »Batmobil« zufrieden, lange vor den automobilen US-Filmhelden. Waren die B.A.T. futuristisch im herkömmlichen Sinne, so ging der Ford FX von 1954 einen Schritt weiter, bediente sich der Motive aus der Jetwelt mit Glaskanzel, düsenähnlichen Frontscheinwerfern, bespickt mit Antennen, und Heckflossen gerieten zu schlanken Leitwerken. Dieses Auto war mehr Show als gangbarer Weg für kommende Modelle. Dies gilt ebenso für den Borgward von 1956. Hier paarten sich Übermut, genährt aus den Erfolgen der Marke, und Mißverständnis von Kundenwünschen der Zukunft. Dabei war die Studie mit Glaskanzel und einem Heckleitwerk ähnlichen Flossen auch Testbasis für neue Motoren für Bundeswehrfahrzeuge mit Boxermotor. Unrühmlich war ihr Ende: Der Wagen landete nach einem Bremsversagen, es waren bereits Scheibenbremsen, an einem Baum.

Als 1955 die Deutschen vom BMW 507 durch seine extreme Eleganz auf Rädern überrascht wurden, mußte es für Raymond Loewy eine Herausforderung gewesen sein, 1957 ein Coupé auf der 507-Basis zu entwerfen. Völlig neue Ideen im Detail verblüfften wegen ihrer machbaren Umsetzung, ohne fremd zu wirken, sei es das asymmetrisch auf der Haube plazierte, in eine Finne eingebettete BMW-Logo, sei es die geschwungene Frontstoßstange als oberer Abschluß der Luftöffnung, sei es die Seitenansicht, die mit blechgeprägter Lichtkante die Kontur eines hinteren Kotflügel nachzeichnete, seien es die Heckleuchten, die in schlanken, horizontalen aus dem Karosseriekörper wachsenden Stummeln saßen. Elegant waren sie, aber für den rauhen Alltag wohl abträglich. Für diesen war auch die Studie Firebird III nicht gedacht. Als Nachfolger von Firebird I und II entwarf Bill Mitchell, GM-Stylingdirektor, dieses Auto 1958 gänzlich chromlos als ultimative Fortsetzung der beiden Vorgänger-Flugzeugmotive auf Rädern. Ähnlichkeiten zur Natur, hier ansatzweise zu Haifischflossen, waren gewollt, um die Biestigkeit des Gefährts zu unterstreichen. Eine formale Anleihe, die Mitchell in vielen späteren Concept-Cars weiter ausbaute. Dieser Firebird schlug auch ein automatisiertes Fahren vor, das mit Hilfe teils im Asphaltbett verlegten Leitkabeln erfolgen sollte.

Wie anders waren da zwei Autos, die der bekannte deutsche Motorjournalist Fritz B. Busch von zwei jungen Designern gestalten ließ: eine Art Van und ein Coupé, um sie 1965 auf der Frankfurter IAA zu zeigen. Michael Conrad und Piu Manzu nutzten für den 340 cm langen Van Technik von Glas, der Marke des Goggomobils. In klaren Flächen von Blech und Fenstern, mit Verzicht auf Chrom und formale Mätzchen auch im Innenraum zeigte der Entwurf, wieviel Platz ein One-Box-Prinzip bieten kann – und was aus diesem Konzept werden kann, verglichen mit dem gedanklichen Serien-Vorläufer Lloyd LT 500 von 1953. Mit dem Coupé auf Basis des NSU Prinz 1000 konnte die Formensprache nachweisen, daß die Dynamik einer Karosserie nicht in Übertreibungen liegt oder Chrom nötig hat, vielmehr in ausgewogenen Proportionen und klarer Flächendisziplin. Die Scheinwerfer ähnelten einem Austin »Frogeye« Sprite, waren aber aus der Not geboren. Es fehlte die Zeit, pünktlich zur IAA Klappscheinwerfer zu installieren. Aus der Not heraus handelte auch VW 1967, was in Kapitel 3 beschrieben ist. Studiert man einzelne Käfernachfolger, dann waren nur wenige Konzepte stilistisch ausgewogen, wie etwa der Porsche-Entwurf für

A classically serene notchback form proves that the less showy a model is, the more long-lived is its acceptance. Unfortunately the Studebaker was not destined for mass production.

No one expected mass production when Bertone revealed the artistic forms of the 1955 B.A.T. 5, the 1956 B.A.T. 7 and the 1957 B.A..T 9 on the chassis of the Alfa Romeo 1900 Super Sprint. While they bordered on the limits of what was tolerable, their elegance avoided the opulence of American examples and had excellent aerodynamic qualities, as could be expected from the round, heavily glazed forms. That function is now performed by plastic component parts when complicated shapes are needed. With Bertone, the Italians reached a pinnacle in the art of metal-working. Incidentally, B.A.T. stands for Berlinetta Aerodinamica Tecnica; most people were also happy to call it a »Batmobile«, long before automotive protagonists featured in the American movies. While the BAT is futuristic in the traditional sense, the 1954 Ford FX went a step further, using motifs from the world of jets – a glass dome roof, jet-like front headlights, a bevy of antennas and slim tailfins. This car was more show than a viable way for upcoming models. The same goes for the 1956 Borgward. The latter was a combination of exuberance fed by the trademark's successes, and a misinterpretation of future customer wishes. At the same time the pilot model with the glass dome and with fins that resembled an empennage was also a test base for new Bundeswehr engines with a boxer engine. It met an inglorious end: After its brakes failed – these were disc brakes – the car crashed into a tree.

When in 1955 the Germans were surprised by the BMW 507 – extreme elegance on wheels – it must have been a challenge for Raymond Loewy to design a BMW-based coupé in 1957. An astounding array of totally novel details had been implemented here, yet without appearing at all strange: the BMW logo asymmetrically placed on the hood and embedded in a fin, or the curved front bumper as the upper termination of the air vent, or the side view that traced the contour of a back fender with a pseudo-edge, or the rear lights that sat in slim, horizontal stubs that grew out of the bodywork. They were certainly elegant, but no doubt inappropriate for the rough and tumble of daily life. The pilot model Firebird III was also not intended for that. Bill Mitchell, the GM head of styling, designed this car in 1958 as a follow-up to Firebird I and II. It had absolutely no chrome and was the ultimate continuation of the two preceding airplane lookalikes on wheels. Similarities to nature, such as shark fins in this case, were intentional, and underline the wild animal nature of the vehicle, a formal borrowing that Mitchell further developed in many later concept

150. Between 1953 and 1955, Bertone with its BAT 5, 7 and 9 demonstrates the supreme art of shaping metal. (Photo: Bertone.)
151. In 1967 VW publishes potential successors to the Beetle: pictured here are four of 30. (Photo: Volkswagen AG.)
152. In 1954 Ford builds the four-wheel rocket FX Atmos. (Photo: Ford.)
153. In 1958 GM builds the four-wheel jet: the Firebird III. (Photo: GM.)
154. The 1965 Autonova Fam has the functional features of the future van. (Photo: Caspers.)
155. The 1956 Borgward Traumwagen is way ahead of its time: too cocky for words. (Photo: Borgward.)
156. In 1957 Raymond Loewy proves that moving away from the usual look makes the BMW 507 elegant. (Photo: Conceptcarz.)

Wolfsburg. Sein Nachteil: der zu laute Unterflurmotor unter der Hinterbank, auch als Heizung für den Innenraum geeignet. Im Vergleich zu Konkurrenzmodellen fehlten vielen VW-Prototypen die Leichtigkeit und Eleganz.

Bertone gelang 1966 mit dem Lamborghini Marzal ein Glanzstück automobiler Form: mit fast voll verglaster Flügeltür, bienenwabenförmigen Motiven überall, im Grill, im Innenraum, dort silbergrau gefärbt, und in Schwarz als raspelförmige Heckfensterblende. Da fällt die Bescheidenheit eines Wartburg 313/2 als Cabrio von 1961 nicht auf, ist aber der Versuch eines an Mangel geübten Landes, das ständig aus den ewig langen Modellzyklen ausbrechen wollte, teilweise mit beachtlichen Lösungen für künftige Trabbis und Wartburgs. Im Westen gab es leichte Wohlstandseinbrüche, als Anfang der 1970er Jahre die sogenannte Ölkrise Sonntagsfahrverbote, hohe Benzinpreise und lauter vernünftige, windoptimierte Autos zur Folge hatte. Das könnte den Entwurf des Opel GT von 1975 beflügelt haben. Glattflächig, klar definiert in Glas und Blech, ist dieser Entwurf mit Schiebetür zeitlos und formal langlebig, ohne langweilig zu sein. Aufregender, aber deshalb auch verwirrender war die Porsche-Studie Panamericana von 1989, eine Mischung aus 911er und Geländewagen, verziert mit schrägen Radausschnitten vorne und hinten, wie die später am Fiat Coupé (Entwurf Chris Bangle) von 1994 imitierten. Dieses Motiv erreichte dankenswerterweise nicht die Serie. Ähnliches darf vom BMW E1 von 1991 gesagt werden. Hier versuchte die neue elektrische Antriebstechnik in Kleinformat den Spagat zwischen Sachlichkeit und BMW-typischem Design. Oder die Sehweise war noch zu wenig eingeübt in Kompaktmodelle, um die aerodynamischen Gestaltungsdetails als ästhetische Lösung zu begreifen. Dies galt auch für den Konkurrenten Audi mit der Studie A2 CC1 1994. Da danken wir der klugen Serienversion des A2 von 1999.

Wie Ford machte auch Mercedes mit dem VRC 1995 den Versuch, aus einem Grundmodell mit verschiedenen Aufsätzen ein Pick-up, ein Coupé oder eine Limousine zu zaubern. Wer aber hat Platz und Lust, zwei von drei Varianten zu bevorraten? Eine Idee ist willkommen, muß aber nicht als Modell im Maßstab 1:1 gebaut werden. Auch ein Toyota Fun Vii 2011 nicht, selbst wenn seine Seiten wie Leinwände bespielbar sind. Da sind Entwürfe wie der Audi Sport Quattro oder ein Cadillac Elmiraj, beide von 2013, wegen ihrer langlebigen Eleganz zu loben. Die wird bei dem Mitsubishi XR-PHEV, auch von 2013, vermißt, dafür versucht der chromverblockte Bug, eine neue

157. Der Lamborghini Marzal von Bertone wurde 1966, was Motive wie die Seitenverglasung und die Verwendung der Bienenwabe für Grill oder Armaturen betrifft, vielen zum Vorbild. (Photo: T. Wood.)

158. 1975 ist der Opel GT2 die Sachlichkeit selbst, verborgen in perfektem Volumen eines sportlichen Autos. Eine Haltung, die so in der Serie nie wiederkam – schade. (Photo: Opel.)

159. 1961 kann der Wartburg 312-3 Sport Prototyp die DDR-Designtalente rehabilitieren. (Photo: Stadtarchiv Eisenach.)

160. Mercedes kommt 1995 mit einer Idee, die Ford in den USA schon viele Jahre vorher kannte: die wandelbare Form. (Photo: Daimler.)

161. 2011 will Toyota das Autokleid zur Leinwand machen. Der Fun Vii ist kein Sicherheitsbeitrag. (Photo: Toyota.)

162. Der Porsche Panamericana ist 1989 kein Ruhmesblatt der Stuttgarter Studios. (Photo: Porsche.)

157. With features as side glazing and the honeycomb look for the grille and fittings Bertone's 1966 Lamborghini Marzal became a model for many other cars . (Photo: T. Wood.)
158. The down-to-earth, functional features of the 1975 Opel GT2 concealed within the perfect body of a sports car – an approach the manufacturer did not repeat in series production – too bad. (Photo: Opel.)
159. The 1961 Wartburg 312-3 Sport Prototype restores the reputation of the GDR's designers. (Photo: Stadtarchiv Eisenach.)
160. In 1995 Mercedes comes up with an idea that the U.S. Ford featured many years earlier: the convertible form. (Photo: Daimler.)
161. The bodywork of the 2011 Toyota is a display panel. This doesn't make the Fun Vii any safer. (Photo: Toyota.)
162. The 1989 Porsche Panamericana does no credit to the Stuttgart studio. (Photo: Porsche.)

cars. This Firebird also proposed automated driving, to be made possible in part by means of cables installed in the asphalt road bed.

How different were the two cars that the famous German automobile journalist Fritz B. Busch commissioned two young designers to create: a kind of van and a coupé that would be exhibited at the 1965 Frankfurt IAA (International Motor Show). For the 340-cm-long van, Michael Conrad and Piu Manzu used Glas technology, the trademark of the Goggomobil. In clear metal surfaces and windows, and without chrome and formal gimmicks in the interior as well, the design showed how much room a one-box-model can provide – and what this concept can lead to, compared with its 1953 conceptual precursor Lloyd LT 500. With the coupé based on the NSU Prinz 1000 the car's style demonstrated that the dynamic nature of bodywork is not due to exaggeration and does not require chrome, but rather consists in well-balanced proportions and a clear control of surfaces. The headlights resembled those of an Austin »Frogeye« Sprite, but were born of sheer necessity: it was too late to install retractable headlights in time for the IAA. In 1967, VW also acted out of necessity, as described in Chapter 3. If we study individual successors of the Beetle, only few concepts were stylistically balanced, for instance, the Porsche design for Wolfsburg. Its disadvantage: the under-floor engine under the back seat, which also heats the interior, is too loud. Many VW prototypes lacked the lightness and elegance of their competitors.

With the 1966 Lamborghini Marzal, Bertone created a gem of automotive form: It had a fully glazed gullwing door, and honeycomb-shaped motifs everywhere – in the grille, in the interior (where they were silver grey), and in black as a grater-like rear window shade. Then the unpretentiousness of a 1961 Wartburg 313/2 convertible is not conspicuous, but is simply the experiment of an East Germany used to scarcity that was constantly trying to break out of eternal cycles of models, in part with impressive solutions for future Trabbis and Wartburgs. In the West there were occasionally slight declines of affluence when in the early 1970s the so-called oil crisis resulted in Sunday driving bans, high gas prices and all kinds of sensible, wind-optimized cars. That probably inspired the 1975 Opel GT. Slipstream, clearly defined in glass and steel, with a sliding door, this design is timeless and formally long-lived without being boring. More exciting, but therefore also more disconcerting, was the 1989 Porsche Panamericana, a blend of the 911 and an all-terrain vehicle, sporting sloping wheel cutouts in front and back, like those later copied on the 1994 Fiat coupé (designed by Chris Bangle). Thankfully this motif was never mass-produced. The same can be said of the 1991 BMW E1. Here, on a small scale, the new electric drive technology attempted to do a balancing act between practicality and a typical BMW design. Or else the designers had had too little practice in compact models to figure out that aerodynamic design details could be an aesthetic solution. This was also true of their competitor Audi with its 1994 trial car A2 CC1. We are grateful for the clever mass-produced version of the A2 in 1999.

Like Ford, Mercedes with the 1995 VRC also made an attempt to conjure up a pickup, coupé or sedan from a basic model, by adding different parts. But who has enough room, or the inclination, to stock two out of three versions? A new idea is welcome, but doesn't have to be built as a model on a scale of 1:1. Not even a 2011 Toyota Fun Vii, even if the color of its sides can be altered by the driver. On the other hand, designs like the Audi Sport Quattro or a Cadillac Elmiraj, both built in 2013, are to be praised because of their long-lived elegance. That is lacking in the Mitsubishi XR-PHEV, also dated 2013; on the other hand, its chrome-blocked front attempts to find a new brand look. The complete opposite of the 2014 Skoda Vision C, this model was the beginning of further studies with extremely sharply defined steel panels and overall balanced proportions. The latter were absent in the 2014 Google car, as though the designers were trying to

Hauslinie zu finden. Das ganze Gegenteil der Skoda Vision C 2014: Dieses Modell war der Beginn weiterer Studien mit extrem scharf definierten Blechflächen und ausgewogenen Proportionen der Gesamtform. Die fehlten beim Google-Auto 2014, als wollten die Designer sagen: Wir bauen keine Autos. Wir liefern nur die Hülle für unsere Software. Eine Hülle der veränderbaren Art hatte Mercedes mit der Studie Concept IAA 2015 präsentiert. Bug und Heck hatten veränderbare Leitflächen zur bereits sehr windschlüpfigen, auf Autobahnen anwendbaren Form. Wer spart, holt sich mit roten, die Kontur fördernden Linien und satiniertem, aber durchsichtigen Glasflächen seriennahe Ideen, die der Opel GT 2016 den Fans leider nicht zum Kauf bot. Ob der Roewe Vision R von 2016 das von sich sagen kann, ist nicht bekannt, als Designexport aus China ist sein Design aber akzeptiert.

Selbst als Studie sind ein Mercedes-Maybach 6 oder der BMW Vision Next 100 von 2016 eher Traumwagen als Konzepte machbarer Zukunft – oder vielleicht Frustkompensation der perfekten Art für auf Großserien geimpfte Designer. Die bewiesen bei VW, daß ein Elektroauto ohne Grill Charakter im Karosseriekleid tragen kann wie beim VW I.D. 2016. 2017 raffte sich BMW auf, seine Hauslinie mit der Studie Concept 8 Series aufzufrischen, ohne Zukunftsvisionen zu sehr zu strapazieren. Alle Genannten sind nur Beispiele einer wachsenden Zahl von Ideen auf Rädern.

163. 2013 kann Üppigkeit noch elegant sein, weil die Form, nicht das Ornament zählt: der Cadillac Elmiraj. (Photo: GM.)

164. Der Mitsubishi XR-PHEV geht 2013 in die Fülle jeglicher Üppigkeit. (Photo: Mitsubishi.)

165. Das Google-Auto von 2014 ohne Design. (Photo: Google.)

166. 2016 zeigt der VW i.d. mögliche Markenidentität. (Photo: Volkswagen AG.)

167. Der Roewe Vision R kann 2016 das, was der Mitsubishi XR-PHEV auch kann. (Photo: Roewe.)

168. 2016 sagt der Mercedes Maybach: Länge läuft. (Photo: Daimler.)

169. Der BMW Vision next 100 macht 2016 das Blech zur elastischen Haut. (Photo: BMW.)

170. 2015 duckt sich der Mercedes Concept IAA unter dem Wind. (Photo: Daimler.)

171. Die rote Linie wird 2016 beim Opel GT nicht überschritten. (Photo: Opel.)

172. 2017 wird der BMW Concept 8 Series grimmig, aber serientauglich. (Photo: BMW.)

163. In 2013 opulence can still be elegant, for it's the form that counts, not the ornamentation: the Cadillac Elmiraj. (Photo: GM.)
164. The 2013 Mitsubishi XR-PHEV – an ultra-luxurious design. (Photo: Mitsubishi.)
165. The no-design 2014 Google car. (Photo: Google.)
166. The 2016 VW I.D. EV Concept. (Photo: Volkswagen AG.)
167. The 2016 Roewe Vision R can do everything the Mitsubishi XR-PHEV can. (Photo: Roewe.)
168. The 2016 Mercedes Maybach: Long cars are faster. (Photo: Daimler.)
169. The 2016 BMW Vision Next 100, where metal morphs into an elastic skin. (Photo: BMW.)
170. The 2015 Mercedes Concept IAA hunkers down under the wind. (Photo: Daimler.)
171. The 2016 Opel GT doesn't cross the red line. (Photo: Opel.)
172. In 2017 the BMW Concept 8 Series starts to look grim, but suitable for series production. (Photo: BMW.)

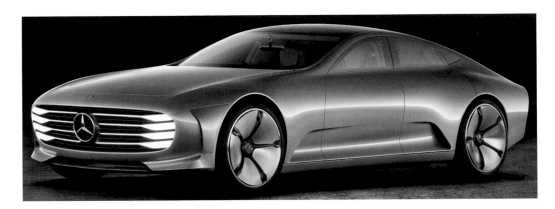

say: We don't build cars. We simply supply the envelope for our software. With the 2015 study Concept IAA, Mercedes presented a transformable shell. Its front and rear had sections that could be modified, adding to an already very aerodynamic form that could be driven on freeways. Less expensive versions had red lines that enhanced the contour und satin-finished but transparent glass panels, ideas that were production-oriented; unfortunately, these were not available to fans of the 2016 Opel GT. It is not known whether the same is true of the 2016 Roewe Vision R. However, this Chinese design model has generally been accepted.

Even as studies, the 2016 Mercedes-Maybach 6 or the BMW Vision Next 100 are dream cars rather than concepts of a feasible future – or perhaps a perfect way for designers who are oriented toward mass production to take out their frustrations. The designers at VW proved that the bodywork of an electric car without a grill can show character, an example being the 2016 VW I.D. In 2017 BMW pulled itself together, smartening up its vehicle line with the study series Concept 8 without overtaxing its visions of the future too much. All the above-named models are merely examples of a growing number of ideas on wheels.

173. Camille Jenatzy brachte 1899 die »Jamais-Contente« auf über 100 km/h – elektrisch. (Photo: Archiv Autor.)

174. 1913 seiner Zeit voraus: Alfa Romeo Ricotti. (Photo: Wikipedia.)

175. Der Rumpler-Tropfenwagen war 1921 windschlüpfiger als Serienautos der 1980er Jahre. (Photo: LoKiLeCh.)

176. 1922 war der Zeiner wie der Alfa Ricotti seiner Zeit voraus – auch im Stoffkleid. (Photo: Archiv Autor.)

177. Die Haube des Skoda Popular 418 neigt sich 1937 vor dem Wind. (Photo: Skoda.)

178. Der Adler 2,5L Sport geht 1938 den windschlüpfigen Weg. (Photo: K. Nahr.)

179. Von Hideo Kodama noch schöner gezeichnet: die Windsbräute der 1920er und 1930er Jahre. (Photo: Kodama.)

6.3. Form und Technik passen sich dem Wind an

Als D-Züge versenkbare Fenster hatten, konnte die flache, horizontal gehaltene Kinderhand Aerodynamik studieren. Die hinten hinuntergedrehte Handfläche wurde nach oben gerissen und vorne nach unten gebogen. Heute hilft ein schnelles Auto mit offenem Fenster. So kann der Einfluß der Luft bei hohem Tempo studiert werden, anwendbar für alle Körper zu Lande und zu Luft. Es dauerte länger, bis sich die Kenntnis durchsetzte, daß der Luftwiderstand ab ca. 60 km/h der größte zu überwindende Widerstand im Fortkommen auf Rädern ist. Intuitiv richtig gemacht waren Formen mit gerundeten oder angespitzten Fronten, auslaufend in röhrenähnlichem Grundkörper, falls es kein kutschengeprägtes Mobil war. 1899 machte der 100 km/h-Überwinder Jenatzy mit seiner »Jamais-Contente« nur die Hälfte richtig. Der raketengleiche Wagenkörper hatte frei stehende Räder, Jenatzy ragte zur Hälfte oben aus der Karosse, weit entfernt von einer anliegenden Umströmung. Näher dran war Graf Marco Ricotti 1913 mit seinem Alfa. Der kompakte Zeppelin auf Rädern hatte lediglich die unverkleideten Räder und den aufgerissenen Bug als Windgegner. Neun Jahre danach machte der Ingenieur Edmund Rumpler noch ähnliche Fehler. Sein Autokonzept mit schiffsähnlichem Grundkörper, gewölbter Dachfläche, ausgerundetem Bug und gerundeter Frontscheibe – quasi ein One-Box-Prinzip – beließ die Kotflügel als Stummelflügel und erkannte den Vorteil eines glatten Unterbodens, hier als geschlossene Wanne ausgebildet. Spätere Nachmessungen im modernen Windkanal ergaben einen Cw-Wert von 0,28. Der Zeiner-Kleinwagen von 1922 wirkte viel windschnittiger. Auch er in Zigarrenform, aber mit völlig in die Gesamtform integrierten Rädern. Sparsamkeit zwang zu textiler Verkleidung, und sparsam war auch sein Auftritt in der Fachliteratur.

Paul Jaray, von Haus aus Flugzeugingenieur, hatte alle Kenntnisse, um dem Auto aerodynamische Manieren beizubringen, wohl wissend, daß Autos anderen Gesetzen des Windes gehorchen als Flugzeuge. Das eine soll tempogetrieben am Boden bleiben, das andere abheben. Als Ergebnis fand er das Prinzip des halben Flügels als Unterbau in der Längsansicht. Darauf setzte er als Aufbau das volle, aber stehende, nach hinten spitz auslaufende Flügelprofil. Wenige Retuschen für formale Übergänge ergaben 1921 die patentierte Karosserie, die danach mehrere Hersteller nutzten – und das in einer Zeit, als die Kutsche noch immer formales Leitbild des Automobils war. Der Fortschritt mit steigender Geschwindigkeit, immer schneller drehenden Motoren, gepaart mit verbesserten Fahrwerken und Bremsen, brachte Vorschläge, wie sich die Karosserie unter dem Wind ducken könnte. Stromlinie wurde zum Reizwort und Stichwortgeber für unzählige Entwürfe, wie den Chrysler Airflow von 1934, den Volvo PV 36 von 1936, den Toyota AA und den Peugeot 402, beide von 1936. 1937 hatte der Skoda Popular ebenfalls diese Bugform. Selbst der Opel Admiral von 1938 trug eine ähnlich ausgerundete Kühlermaske wie die genannten Vorbilder. Kon-

6.3. Form and technology adapt to the wind

Back when express trains had windows that could be lowered, a child's flat, horizontally held hand could study aerodynamics. Your palm, which was facing backwards, was abruptly jerked up and bent downward in front. Today a fast car with its window open fulfills the same function. This is how the effect of air at high speed can be studied, and the result applied to all solids on land and in the air. It took longer for people to learn that air resistance starting at roughly 60 km/h is the greatest resistance that must be overcome when one travels on wheels. Engineers intuitively correctly designed forms with rounded or pointed front sections, ending in a basic tube-like body, unless it was a vehicle modeled on a carriage. In 1899 the 100-km/h record breaker Jenatzy with his »Jamais-Contente« got it only half right. The rocket-like body of the car had free-standing wheels, and Jenatzy's upper body stuck halfway out of the car, a far cry from laminar flow. Count Marco Ricotti with his Alfa in 1913 got closer. On this compact Zeppelin on wheels, only the naked wheels and the wide-open front section resisted the wind. Nine years later the engineer Edmund Rumpler was still making similar mistakes. His automobile concept with a ship-like basic body, curved roof, rounded front end and rounded windshield – practically a one-box configuration – retained the fenders as stubby wings and realized the advantages of a smooth underbody, here shaped like a closed pan. Later repetitive measurements in a modern wind tunnel produced a drag coefficient of 0.28. The 1922 Zeiner Kleinwagen (minicar) seemed to be much more aerodynamic. It, too, was cigar-shaped, but had wheels that were completely integrated in the overall shape. For the sake of economy the car was fabric-clad. It was rarely mentioned in the specialized literature.

Paul Jaray, originally an aircraft engineer, knew all about aerodynamics, and was well aware that cars obey different wind laws than do aircraft. A car propelled by speed is meant to stay on the ground, while aircraft is meant to lift off. As a result he discovered the principle of the half-spindle profile as the chassis. On it he set the body – the full but standing wing profile that tapers in the back. From a formal point of view the patented body was slightly retouched in 1921 and then used by several manufacturers – and that in an era when the form of the automobile was still modeled on the carriage. Progress came with increasing speed, faster rotating motors in con-

173. In 1899, Camille Jenatzy produces the electric car »Jamais-Contente«, which reached over 100 km/h. (Photo: author's archive.)
174. Ahead of its time in 1913: Alfa Romeo Ricotti. (Photo: Wikipedia.)
175. The 1921 Rumpler Tropfenwagen was more streamlined than 1980s series cars. (Photo: Lo-KiLeCh.)
176. The 1922 Zeiner and Ricotti's Alfa were ahead of their time – even with a body made of fabric. (Photo: author's archive.)
177. The hood of the 1937 Skoda Popular 418 leans away from the wind. (Photo: Skoda.)
178. The 1938 Adler 2.5-litre Sport goes streamlined. (Photo: K. Nahr.)
179. Even lovelier as drawn by Hideo Kodama: the whirlwind models of the 1920s and 1930s. (Photo: Kodama.)

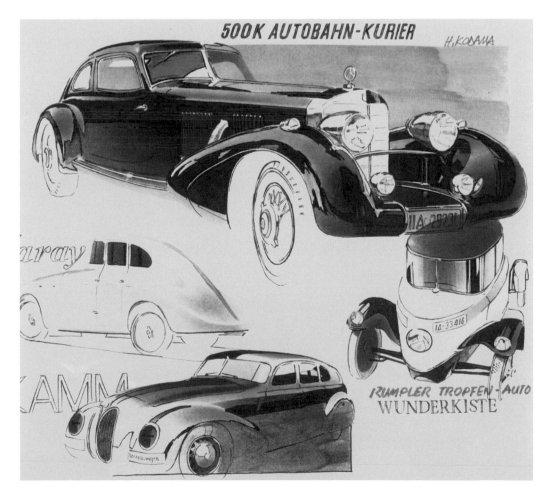

sequent in der Erkenntnis windgerechter Formgebung folgte der Adler 2,5 Liter Sport von 1938 jenen Vorgaben mit abfallendem Rundbug und abfallendem Heck. Der kleine Österreicher Steyr 55 mit dem Gesicht eines Boxerhundes machte es 1938 genauso, der KdF-Wagen und Nachkriegskäfer von Ferdinand Porsche gleichfalls. Da es noch keine Windkanäle für Automobile gab, behalf man sich vor dem Kriege mit Prototypen, auf die Wollfäden in geregelten Abständen geklebt wurden. Das fahrende Auto wurde von einem Begleitwagen photographiert oder gefilmt, um in der Bildauswertung herauszufinden, ob diese Fäden ungestört an der Blechhaut anlagen oder verwirbelten. Letzteres war der Hinweis auf eine nicht mehr anliegende Strömung, was eine Verschlechterung des Luftwiderstands bedeutete. Ist der erst einmal optimiert, läßt sich erheblich Benzin sparen, läßt er sich schneller fahren oder ersatzweise ein kleinerer Motor verwenden. Der Maybach SW 38 Stromlinie von 1938 zeigte das Jaray-Prinzip und die Technikhilfe mit den Wollfäden. Neumann-Neanders Fahrmaschine, welch schöner Name, mußte 1939 mit beschränkten Mitteln versuchen, mit knappem Hubraum, aber windschlüpfiger Form flott voranzukommen. Formal eher selten in dieser Fahrzeugklasse!

Die Marke Tatra hatte ihr mutiges 1930er-Jahre-Konzept über den Krieg gerettet: heckgetrieben und konsequent im Aerodynamikdesign. Lufthutzen waren so windschnittig konzipiert wie die angedeutete Panorama-Frontscheibe. Erst das letzte Modell (Entwurf von Vignale) von 1973, weiterhin mit Heckmotor, verlor sich im üblichen Design dieser Zeit. Völlig unüblich und schnell zur Ikone modernen Automobildesigns wurde der Citroën ID/DS von 1955. Der Bildhauer Bertoni hatte mit dieser Autoarchitektur nicht nur skulpturale, sondern auch aerodynamische Bestleistung erbracht, gefördert durch den flachen, im Grundriß gerundeten Bug mit ebenso rundlich ausgelegter Frontscheibe, glatten Seitenflächen, verkleideten Hinterrädern und einem passablen Heckabschluß ohne starke Verwirbelung in der sogenannten Totwasserzone. Das war jene Zeit, als Bestwerte bei dem Cw-Wert nur möglich waren, wenn das Auto ganz auf klassische Konturen verzichtete und eher einer idealen, physikalisch definierten Aerodynamikform folgte. Solch ein Beipiel war die von Pininfarina 1978 vorgestellte »Forma aerodinamica ideale«. Mit dem Ziel, den Innenraum eines Fiat 130 zu bieten, hätte sie serienfertig einen Cw-Wert von ca. 0,23, als Massenmodell kam sie auf beachtliche 0,175. Ganz so niedrig schaffte es der Audi 100 nicht. Sein Modell von 1983 war straßenfertig gut für 0,30, möglich dank rundum extrem flächenbündiger Scheiben – eine Kunst bei versenkbarer Ausführung. Halbverkleidete Hinterräder, minimale Fugenbreiten, extrem kontrollierter Luftdurchsatz im Innenraum und viele kleine Helferchen ermöglichten den weltmeisterlichen Wert. Wenig gönnerhafte Naturen bemäkelten die schräg stehenden Seitenfenster und großen Glasflächen, ein Angriff der Sonnenstrahlen auf die Innenraumtemperatur, der sich bald mit serienmäßig eingebauten Klimaanlagen legte.

Das war nicht das Ende machbarer Windschlüpfigkeit für Stufenheckmodelle. Schon der NSU Ro 80 konnte dank kleinem Wankelmotor mit niedrigem Bug und hohem Heck sehr gut unter dem Wind fahren. Dieses Konzept war nur der Anfang. Immer höher stieg die Oberkante des Kofferraums, und selbst Hubkolbenmotoren im Bug konnte man tiefer legen mittels geneigtem Einbau. Egal ob ein VW Jetta oder ein Alfa Romeo 164, fast jede Stufenheckversion übernahm das Rezept. Der Wettlauf zur Überwindung des Luftwiderstands wurde zum Dauerlauf. Der vielzitierte Cw-Wert ist nur ein Teil des Luftwiderstands, man achtete genauso auf die sogenannte Querschnittsfläche eines Modells. Das ist die Umrißfläche quer zur Längsrichtung an der Stelle, wo sie am größten ist. Das Produkt aus dem Cw-Wert und der Querschnittsfläche ergibt den Luftwiderstand einer Karosse. Diese Berechnungsart zeitigt eigenartige Ergebnisse: So hat, man glaubt

180. 1983 sieht man dem Audi 100 nicht an, daß er nichts vom Luftwiderstand hält; er reduziert ihn auf das absolut Notwendige und seinerzeit Machbare. (Photo: Audi.)
181. Ein Maybach SW38 »Stromlinie« kann 1938 auch die Jaray-Prinzipien tragen. (Photo: Daimler.)
182. 1939 machte sich die kleine Fahrmaschine von Neumann-Neander ganz windschnittig. (Photo: Prototypen-Museum.)
183. Einer macht 1955 alles anders: Form, Bauteilauflösung, Hydraulik als Federung, windfreundlich, leichtes Reisen: Der Citroën ID/DS wird zur Ikone für immer. (Photo: Citroën.)

180. From looking at the 1983 Audi 100 you can't tell that it doesn't care about drag: it reduces it to what is strictly necessary and possible at the time. (Photo: Audi.)

181. The design of a streamlined 1938 Maybach SW38 may also be based on the Jaray principles. (Photo: Daimler.)

182. In 1939 Neumann-Neander's little driving machine was quite streamlined. (Photo: Prototypen-Museum.)

183. The one who did everything differently in 1955: form, component parts, hydraulics suspension, streamlining, easy travel: The Citroën ID/DS becomes an eternal icon. (Photo: Citroën.)

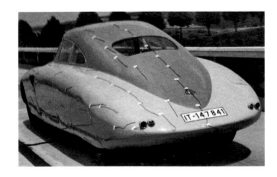

junction with improved chassis and brakes, bringing with it proposals as to how the body could duck below the wind. Streamline became the motto of the day and the keyword for countless designs, like the 1934 Chrysler Airflow, the 1936 Volvo PV 36, and the Toyota AA and the Peugeot 402, both built in 1936. In 1937 the front section of the Skoda Popular also had this shape. Even the 1938 Opel Admiral sported a rounded radiator grill similar to that of the above models. Consistent in acknowledging aerodynamic design principles, the 1938 Adler 2.5 Liter Sport incorporated a sloping round front end and a sloping tail end. The little Austrian Steyr 55 with the face of a boxer dog followed suit in 1938, and so did Ferdinand Porsche's Nazi-era and postwar Beetle. Since wind tunnels for cars did not exist yet, prewar designers made do with prototypes on which woolen threads had been glued at regular intervals. The moving car was photographed or filmed by a vehicle driving alongside it: Image analysis would then show whether those threads lay undisturbed close to the sheet metal body or swirled away from it. The latter was an indication that the airflow is no longer snug, which means worsening drag. Once drag has been optimized, the car uses considerably less gasoline, and can be driven at faster speeds or, alternatively, it is possible to use a smaller motor. The 1938 Maybach SW 38 streamline car used the Jaray principle and the woolen thread technique. In 1938, with limited funds, Neumann-Neander's Fahrmaschine (driving machine) – what a great name – had to try and get along with limited engine capacity, but with an aerodynamic shape. From a formal standpoint it was somewhat of a rarity in this class of vehicles!

The Tatra brand had maintained its courageous 1930s concept beyond World War II: rear-wheel-driven and consistent in its use of an aerodynamic design. The concept of its air scoops was as aerodynamic as its suggestion of a wrap-around windshield. Only the last model (designed by Vignale) in 1973, still with a rear-mounted engine, got lost in the typical design of this period. The totally untypical 1955 Citroën ID/DS quickly became the icon of modern automobile design. This architectural feat was not only the sculptural but also the aerodynamic personal best of the sculptor Bertoni, enhanced by the rounded front end, with an equally rounded windshield, smooth side panels, cowled rear wheels and a passable rear end without strong turbulence in the so-called dead-water zone. This was the era when optimum drag coefficients were possible only if the car completely dispensed with classic contours and instead followed an ideal, physically defined aerodynamic form. An example of this was the »Forma aerodinamica ideale« introduced in 1978 by Pininfarina. Intended to have the interior of a Fiat 130, it would have had a drag coefficient of roughly 0.23 when ready to go into production, while as a mass model it attained an impressive 0.175. The Audi 100 did not manage to bring down the drag efficient quite as low. Its 1983 model, when ready for the road, was good for 0.30, which was possible thanks to all-around flush-mounted windows – quite a trick with a retractable style. Half cowled rear wheels, minimal gap widths, a highly monitored airflow rate in the interior and many small devices made the car a world champion. More invidious persons found fault with the sloping side windows and large glass surfaces, as a result of which the interior heats up in the sun, a defect that was soon remedied with mass-produced built-in air-conditioning.

This was not the end of feasible aerodynamic notchback models. Even the NSU Ro 80 was easily able to drive under the wind thanks to a small Wankel engine with a low front and a high rear section. This concept was only the beginning. The upper edge of the trunk kept getting higher, and even reciprocating engines in the front section could be positioned lower down if in-

es kaum, ein Formel-1-Renner einen miserablen Cw-Wert zwischen 1,1 und 1,2. Nur die winzige Querschnittsfläche macht ihn am Wind günstiger. Optisch lächerlich erscheinen besonders an den Frontflügeln die vielen kleinen Stege in unterschiedlichsten Verformungen, damit geringster Windhauch in die gewünschte Richtung weht. Unsichtbar oder wenig erkennbar bleiben in der Serie ähnliche Hilfsmittelchen, sei es an Radausschnitten, seien es scharfkantige Flächenabschlüsse an der C-Säule, seien es glatte Flächen am Unterboden oder gezahnte Bleche vor den Rädern, einem Spachtel nicht unähnlich. Und oft gönnen sich PS-starke Versionen einen ausgeprägten Heckdiffusor, das Hilfsmittel zur kontrollierten Abführung des Windes unter dem Auto. Die Beispiele von Mercedes geben, wie viele Hilfsmittel, auch optimierte Cw-Werte an. Dank moderner Windkanäle kann noch an der dritten Kommastelle eines Cw-Wertes gefeilt werden.

Zum Schluß etwas Physik in Formelform. Das Schaubild nennt alle Einflüsse zur Ermittlung des Luftwiderstands. Der Cw-Wert aber läßt sich nur im Windkanal-Versuch ermitteln, weil die aerodynamische Güte einer Form lediglich hier gemessen werden kann. Der Kanal kann heute Wetter- und Temperaturunterschiede simulieren. Am Ende dürfte die Loewy-Schautafel zum Thema Aerodynamik eher ironisch zu lesen sein. Letzte Bitte: Sagen Sie bitte nie »windschlüpfrig«.

184. 2015 bemüht sich ein Mercedes GLC, daß die Rauchfäden im Windkanal schön gekämmt bleiben. (Photo: Daimler.)

185. Doch wie es darunter aussieht, geht den Aerodynamiker etwas an, denn da sitzen Reserven für gute Cw-Werte. (Photo: Daimler.)

186. Zum Nachrechnen, wenn einer den Cw-Wert ermitteln will. (Photo: Archiv Autor.)

187. Der Windkanal macht viel Wind, ist deshalb so groß, siehe den kleinen roten Benz mittendrin. (Photo: Daimler.)

188. Raymond Loewy, amerikanisches Supertalent im Design, ahnt den Einfluß des Windes auf alles. (Zeichnung: Loewy.)

Verbesserte Kühlerumfeldabdichtung und Strömungsführung zur effizienten Ausnutzung der Kühlluft

Klassenbester c_w-Wert von 0,31

Aeroakustische und aerodynamische Gestaltung von A-Säule und Außenspiegel

Verlängerter Dachspoiler mit optimierten seitlichen Abrisskanten

Scheinwerferumfeldabdichtung

Strömungsgünstige Gestaltung der Bug- und Heckschürze

Dreidimensionaler Radspoiler mit patentierten Radlaufschalen zur optimierten Radumströmung

Optimierte Unterbodenströmung mit großflächigen Motorraum- und Hauptbodenverkleidungen

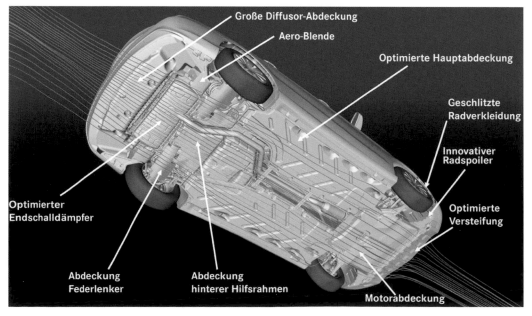

Große Diffusor-Abdeckung

Aero-Blende

Optimierte Hauptabdeckung

Geschlitzte Radverkleidung

Innovativer Radspoiler

Optimierte Versteifung

Optimierter Endschalldämpfer

Abdeckung Federlenker

Abdeckung hinterer Hilfsrahmen

Motorabdeckung

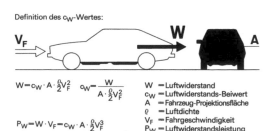

Definition des c_w-Wertes:

$$W = c_W \cdot A \cdot \frac{\varrho}{2} V_F^2 \qquad c_W = \frac{W}{A \cdot \frac{\varrho}{2} V_F^2}$$

$$P_W = W \cdot V_F = c_W \cdot A \cdot \frac{\varrho}{2} V_F^3$$

W = Luftwiderstand
c_W = Luftwiderstands-Beiwert
A = Fahrzeug-Projektionsfläche
ϱ = Luftdichte
V_F = Fahrgeschwindigkeit
P_W = Luftwiderstandsleistung

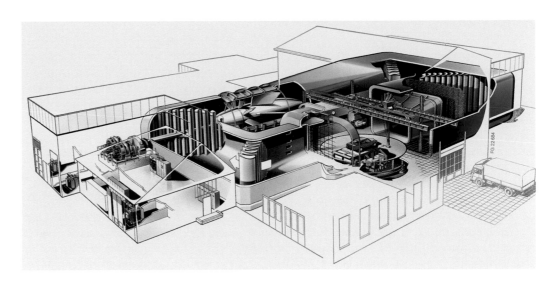

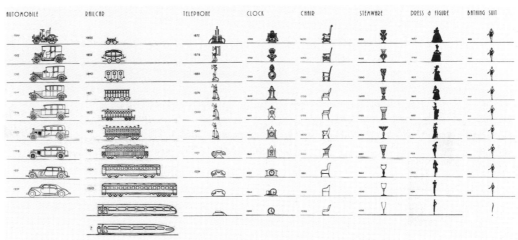

stalled at a slant. Regardless of whether it was a VW Jetta or an Alfa Romeo 164, almost every notchback version adopted this recipe. The race to overcome air resistance turned into an endurance run. The often-cited drag coefficient is only a fraction of air resistance; another factor that was considered was the so-called cross-sectional area of a model. This is the contour area transverse to the longitudinal direction at the place where it is greatest. The product of the drag coefficient and the cross-sectional area is the air resistance of a car body. This calculation method leads to curious results: Thus it is hard to believe that a Formula One race car has a miserable drag coefficient of between 1.1 and 1.2. Only its infinitesimal cross-sectional area makes it more advantageous in terms of wind resistance. Especially ridiculous from a visual point of view are the many small deflectors, particularly on the front wings, in all sorts of deformed shapes, so that the slightest breeze will blow in the desired direction. Invisible or hardly recognizable are similar minor aids in the series cars – on wheel cutouts, or sharp-edged terminations on the C pillar, or smooth panels on the underbody, or serrated metal parts in front of the wheels, not unlike spatulas. Often powerful models add a distinctive rear diffuser, which helps dissipate the wind under the car in a controlled way. The examples of Mercedes even specify optimized drag coefficients. Thanks to modern wind tunnels, drag coefficients can be determined down to the third decimal place.

Finally here are some physics in the form of formulas. The chart lists all factors involved in determining air resistance. However, the drag coefficient can only be determined in an experiment in the wind tunnel, because the aerodynamic quality of a shape can be measured only here. Nowadays the tunnel is capable of simulating weather and temperature differences. In the end, the Loewy chart on the topic of aerodynamics might have to be read with tongue in cheek. A last request: Never use the word »windschlüpfrig« for »streamlined« in German. The correct word is »windschlüpfig«.

184. The 2015 Mercedes GLC trying to keep the threads of smoke in the wind channel nice and straight. (Photo: Daimler.)
185. Aerodynamics experts care what things look like underneath, for that ensures good drag coefficients. (Photo: Daimler.)
186. Here's how you calculate the drag coefficient. (Photo: author's archive.)
187. The wind channel creates a lot of wind, which is why it's so large. For comparison, look at the little red Benz. (Photo: Daimler.)
188. Raymond Loewy, American star designer, senses that the wind influences everything. (Drawing: Loewy.)

6.4. Die Form hilft der Technik bei der Sicherheit

Das Leben mit dem Auto folgt individuellem Verhalten. So sind Unfälle mit dem Auto individuelle Ereignisse. Kommt daher unsere Verweigerung, die Konsequenzen von Unfällen so wahrzunehmen wie Flugzeugkatastrophen oder Bahnunglücke? Erschreckend ist selbst heute die Zahl der Unfallopfer im Straßenverkehr: EU-weit sind es immer noch 25 000 pro Jahr. Der starke Anstieg des Autobestands in den Wohlstandsländern, aber mit Verzögerung ebenso in der dritten Welt, fordert jährlich Todesopfer horrenden Ausmaßes. Das Beispiel Deutschland allein reicht aus, dies darzustellen. 1912 gab es rund 70 000 Automobile im Land, aber je 100 000 Einwohner starben 631 Personen bei Unfällen. Verbesserte Technik, verbesserte Straßen, besserer Umgang mit dem neuen Transportmittel mögen Gründe dafür gewesen sein, daß 1938 bei einem Bestand von 3,2 Millionen Fahrzeugen trotzdem immer noch 230 Personen je 100 000 Einwohner tödlich verunglückten. Wie sehr diese Tragik dank verbesserter aktiver und passiver Sicherheit, dank niedriger Promillegrenzwerte, dank verbesserter Rettungssysteme, gemildert werden kann, erklären die aktuellen Zahlen von 2015: Bei einem Bestand von 53,7 Millionen Fahrzeugen starben nur noch sechs Personen je 100 000 Einwohner im Verkehr. Dabei hat sich die jährliche Fahrleistung pro Person in den Jahrzehnten erheblich gesteigert. Dies ist der technischen Entwicklung zu verdanken, weil die meisten Unfälle auf menschlichem Versagen beruhen. Wesentliche Ursachen dafür sind zu hohe Geschwindigkeiten und Alkoholeinfluß.

Deshalb nutzen Ausländer gern unsere Autobahnen zum Austesten des Vollgases. Hilft dagegen ein Tempolimit? Nicht viel. Die USA haben vergleichsweise mehr Verkehrstote. Technisches Versagen kann dank des hohen Standards statistisch vernachlässigt werden. Dieser wurde seit den 1950er Jahren intensiv bearbeitet und hatte Anfang der 1970er Jahre neue Sicherheitskonzeptfahrzeuge zu bieten, zeitgleich mit den höchsten jährlichen Todesfällen, das war 1970 mit 19 193 Toten. 2016 waren es dankenswerterweise nur noch 3214.

Blickt man zurück, so haben mehr skurrile als sinnvolle Vorschläge zur verbesserten Sicherheit beigetragen. Autos waren extrem unverformbare Konstruktionen. Im Crash blieb das steife Fahrgestell wenig beeinträchtigt. Holzrahmenkonstruktionen gingen zu Bruch, Beleuchtungssysteme waren Grund für Brände, nichts war bewußt für die Sicherheit ausgelegt, vergessen waren jene Technikverbesserungen für sichereres Vorankommen. Ein Beispiel für skurrile Einzelversuche, die Reise abzusichern, war 1916 der Vorschlag der Marke Protos. Zur Vermeidung von Reifenpannen baute man Räder, die zwischen Nabe und Felgenkranz Federn hatten, um die Luft zur Glättung unebener Straßen zu ersetzen. Nicht überliefert ist die Kurvenstabilität, wie Seitenkräfte auf diese Konstruktion wirkten. Mit der Einführung selbsttragender Karosserien zeigte sich, daß diese Bauart besser verformbar war als alle anderen zuvor. Vielleicht hatte das dazu beigetragen, Unfallfolgen abzumildern. Einen anderen skurrilen Einzelversuch, diesmal das gesamte System Automobil betreffend, führte der amerikanische Pater Alfredo Juliano mit seiner »Aurora« durch. Ihn mögen die Schrecken der Verkehrsopfer angetrieben haben, 1957 ein Auto vorzustellen, das auch an Verkehrsteilnehmer außerhalb der Fahrgastzelle dachte. Der schaufelförmige, weich gerundete Bug sollte Fußgänger auffangen. Innen glänzten die zusammenschiebbare Lenksäule, Sicherheitsgurt, Überrollkäfig und gepolstertes Armaturenbrett. Es blieb eine Solonummer mit der Moralpredigt, daß die Autoindustrie Sicherheitsaspekte bisher verschlafen hatte. Eine der ersten Marken in Sachen Sicherheit war Volvo. Hier wurde ab 1959 der Dreipunkt-Sicherheitsgurt serienmäßig verbaut, wo andere Marken glaubten, ein schüsselförmiges Lenkrad mit gekürzter Lenk-

189. 1972 kann Opel einem Safety OSV noch passable Formen abgewinnen, die seriennah sind. (Photo: Opel.)

190. Die Luft ist raus, Federn sind 1916 beim Auto der Marke Proto Plattfußvermeider. (Photo: Just-a-carguy.)

191. 1957 glaubte der Pater Juliano aus den USA, der Menschheit mit seiner »Aurora« helfen zu können. (Photo: A. Saunders.)

192. 1972: Fiat sollte sich schämen, den ESV 179 so im Land der schönen Autos zu zeigen. (Photo: Fiat.)

193. Volvo machte 1959 den Vorreiter mit serienmäßigen Gurten im Amazon-Modell. (Photo: Volvo.)

189. Opel's 1972 Safety Vehicle OSV still has reasonable, close-to-production forms. (Photo: Opel.)
190. Things have gone flat: Springs, in 1916, help the Proto brand avoid flat feet. (Photo: Justacarguy.)
191. In 1957 Father Juliano thought he could help humanity with his American safety vehicle, the »Aurora«. (Photo: A. Saunders.)
192. 1972: Fiat ought to be ashamed showing the ESV 179 in the country of handsome cars. (Photo: Fiat.)
193. Volvo, in 1959, was the forerunner, with standard seat belts in its Amazon model. (Photo: Volvo.)

6.4. The form helps technology improve safety

Life with a car is determined by individual behavior. Thus car accidents are individual events. Is this why we refuse to look at the consequences of auto accidents in the same way as we look at airplane accidents or train wrecks? Even today the number of traffic accident victims is staggering: in the EU, 25 000 people per year are still killed in car crashes. Every year the large rise in the number of cars in the wealthy countries, but increasingly in the Third World as well, claims fatalities on a horrendous scale. The example of Germany alone is sufficient to illustrate this. In 1912 Germany had 70 000 cars, but 631 persons per 100 000 inhabitants died in accidents. Improved technology, improved roads, more familiarity with the new means of transport may have been the reasons why in 1938, with a total number of 3.2 million vehicles, 230 out of 100 000 inhabitants were killed in traffic accidents. The fact that this tragic figure can be reduced thanks to improved active and passive safety, thanks to low alcohol limits, thanks to improved emergency services, is demonstrated by the current 2015 figures: With a total number of 53.7 million vehicles on the road, only six out of 100 000 inhabitants died in traffic-related accidents. At the same time the annual distance driven per person has considerably increased in recent decades. We owe this to technical development, for most accidents are caused by human error. Important causes of accidents are excessive speeds and driving under the influence of alcohol.

That is why foreigners like to use our German freeways in order to test their car at full throttle. Would a speed limit help? Not much. The U.S. has comparatively more traffic fatalities. Thanks to the high standard of technology, technical failure is statistically negligible. Since the 1950s this standard has been intensively addressed, and in the early 1950s new safety concept cars became available at the same time as there was the highest annual fatality rate; this was in 1970 with 19,193 dead. In 2016, thankfully, the number was down to 3,214.

In retrospect, suggestions that are more bizarre than practical have contributed to improved safety. Cars used to be extremely non-deformable constructions. In a crash the rigid chassis was largely unaffected. Wood frame constructions broke into pieces, lighting systems caused fires, nothing was intentionally designed to improve safety, and designers had forgotten to make technical improvements that would ensure safer travel. An example of a quirky attempt to enhance travel safety was a proposal, in 1916, by Protos. To avoid flat tires they built wheels that had springs between the hub and the rim to take the place of the air that would have ensured a smooth ride on uneven roads. There is no record of the car's curve stability, or how lateral forces affected this construction. When unitized bodies were introduced, it turned out that this version was more easily deformable than all the previous ones. Perhaps this had contributed to reducing the consequences of accidents. The American Fr. Alfredo Juliano carried out another outlandish experiment, this time involving the entire automotive system, with his »Aurora«. The horror of traffic fatalities may have impelled him in 1957 to introduce a car that had in mind road users outside the passenger compartment as well. The shovel-shaped, softly rounded front of the car was supposed to catch pedestrians. Gleaming inside were the collapsible steering column, safety belt, roll cage and upholstered dashboard. The car was to remain one of a kind, and its moral was that the auto industry up to that point had missed the boat when it came to safety products. One of the first trademarks that addressed safety issues was Volvo. Starting in 1959, their vehicles had standard three-point safety belts, where other brands believed a key-shaped steering wheel with a shortened steering column was the ultimate in safety. The big exception was Mercedes. Here chief engineer Béla Barényi, who had patents on the complete car body, held the view that the front and rear of the car should be more easily deformable to reduce the force of the impact, while the passenger compartment should be kept extremely stable. As a standard this was first introduced in 1959 in the W111, the »Heckflossen (Tailfin) Benz«. Today it is a worldwide standard. When, in the early 1970s, America made the raising of safety standards statutory, ugly log-like bumpers were used to absorb the force of the impact. At the same time new safety cars were in the planning. Fiat's treatment of the issue was completely excessive. Here massive rubber front and rear bumper bars were installed on the bodywork of small cars – a mistaken inspiration based on the bumper cars at fairs. In Nissan cars, a roof periscope replaced the side mirrors, as though the latter posed more risk of injury. Opel's OSV was reminiscent of the Kadett and used the design idea of the Vauxhall cars: the front part was tilted backwards to soften the impact of collisions – and it also made the car more aerodynamic. Volvo's safety version had a similar concept, where the oversized bumper system spoiled everything. But Volvo must not have been troubled by this distorted feature; like Mercedes, they had realized early on that auto safety was a much-appreciated selling point. The ESF versions of Mercedes were an acceptable mixture of de-

säule sei das Nonplusultra an Sicherheit. Die große Ausnahme war Mercedes. Hier vertrat der führende Ingenieur Béla Barényi mit Patenten bei der Komplettkarosse die Ansicht, daß Bug und Heck zum Abbau der Aufprallenenergie leichter zu verformen sein sollten, während die Fahrgastzelle extrem stabil gehalten wurde. Serienmäßig wurde dies erstmalig 1959 beim W111, dem »Heckflossen-Benz«, eingeführt. Heute ist das weltweiter Standard. Als Amerika Anfang der 1970er Jahre die Sicherheitsstandards gesetzlich anhob, kamen häßliche Stoßstangenbalken zum Einsatz, um vorgeschriebene Aufprallkräfte zu schlucken. Gleichzeitig plante man neue Sicherheitsautos. Ihnen war das Schutzbedürfnis anzusehen. Formal völlig überzogen behandelte Fiat das Thema. Hier mußten die Karossen der Kleinen herhalten, um massige Bug- und Heck-Gummiprofile aufzunehmen – eine falsch verstandene Inspiration von Jahrmarkt-Scootern. Bei Nissan ersetzte ein Dachperiskop die Seitenspiegel, als ob die ein größeres Verletzungsrisiko seien. Bei Opel hatte der OSV Anklänge an den Kadett und nutzte die Designidee der Vauxhall-Schwestern: die schräg nach hinten geneigte Frontpartie zur Milderung bei Kollisionen – und der Windschlüpfigkeit half es auch. Volvo hatte mit seiner Sicherheitsversion ein ähnliches Konzept, bei dem die überdimensionierte Prallfläche alles verdarb. Aber Volvo mußte diese Verrenkung nicht provozieren, diese Marke hatte, ähnlich wie Mercedes, schon frühzeitig die Sicherheit im und am Auto als geschätztes Verkaufsargument erkannt. Die ESF-Versionen von Mercedes waren eine akzeptable Mischung von Design und Sicherheit, für jeden erkennbar durch gummibewehrte Partien, die bei Unfällen besonders betroffen sind. Das entsprechende Schaubild belegt die umfangreichen Arbeiten zu diesem Thema. Allein die Einführung von Rückhaltesystemen wie Gurt und Airbag stellte die Designer vor neue Herausforderungen im Innenraum. Aus spindeldünnen Fensterpfosten der A-, B- und C-Säule wurden dicke Pfosten aus hochfesten Stählen für gesteigerte Crash-Stabilität. Zusätzlich aufkommende Airbags in den Seitenbereichen und in den Pfosten schränkten die frühere großzügige Rundumsicht ein. Die vielen genannten Beispiele von Sicherheitsautos stammen von 1972. Zum besseren Verständnis wird der englische Begriff des »safety cars« nicht benutzt, der gehört zum Formel-1-Einsatz, wenn es kracht und Boliden im Zaum zu halten sind. Ein zweiter Anlauf, das Auto noch sicherer zu machen, war die Aktion Auto 2000 aus dem Jahre 1981. Das VW-Beispiel war ein designmäßiger Vorläufer des darauffolgenden Passats, der sich trotz Frontmotor ein Gesicht ohne Grill gab. Da war die ESVW-Version von 1972, die Mischung aus VW-Scirocco und Ro 80, schon gefälliger, trotz dicker Gummiapplikationen und kräftiger Stoßstangen. Das Auto 2000 von Mercedes versuchte sich in einer flachen, aber unbeholfenen Version der gestreckten S-Klasse. Aerodynamisch nutzte das Konzept eine riesige Glaskanzel mit Abrißkante im Heck; aus der Stufenhecklimousine wurde dadurch ein Schrägheckauto. Aber so ist das mit der Medizin: Diejenige, die schmeckt, wird als weniger heilsam empfunden. Es bleibt eine offene Frage, ob das auch für viele Sicherheitskonzepte für Autos gilt. Bei dem Thema Sicherheit darf den Designern und natürlich auch den Ingenieuren, deren Arbeit für das ungeübte Auge eher verborgen bleibt, bescheinigt werden, daß aus unbeholfenen Sicherheitskabinen Autos geworden sind, die diskret unfallmindernde Bauteile anbieten, was psychologisch positiv zu werten ist. Denn wer ständig durch offensichtliche Einrichtungen daran erinnert wird, wie gefährlich noch heute der Straßenverkehr ist, wird die Begeisterung für das Auto anders einschätzen. Zum Nachdenken gilt: Das Sterberisiko im Auto ist immer noch mehr als 100mal höher als beim Fliegen.

Die aktive und passive Sicherheit als Standard ist etabliert, selbst bei vielen elektronischen Hilfsmitteln zum Einhalten abgesicherter Fahrzustände. Der Umgang mit kontrollierfähigen Hilfsmitteln ist ein neuer Prozeß, den der Fahrer lernen muß, richtig zu bedienen. Navigationsgeräte vermindern unsichere Suchfahrten. Neue Lichtsysteme mit variantenreichen, sensorgesteuerten Belichtungsmodi berücksichtigen den Gegenverkehr und verbessern Nachtfahrten. Die Designer bekommen gleichzeitig neue Gestaltungsmöglichkeiten bis hin zu überzogenen Lichtgraphiken an Bug und Heck. Spannender werden Bedienungstechniken, wenn sie, von Gesten gesteuert, Befehle übertragen müssen. Es entfallen Schalter, die einst für Verletzungen verantwortlich waren, Gestensteuerung aber muß hundertprozentig verläßlich arbeiten, speziell bei Fahrbefehlen. Wer reagiert in Panik schon regelkonform zu den vorgegebenen Gesten?

194. VW kann 1971 ein wenig Eleganz in den ESVW hauchen, die der Serie schon gutgetan hätte. (Photo: Volkswagen AG.)
195. Mercedes gab sich 1981 keine große Mühe, Sicherheit und Windschlüpfigkeit zu kombinieren. (Photo: Daimler.)
196. Daten des Mercedes ESF13 von 1972.
197. Mercedes hatte 1972 beim ESF13 das Gummi behutsam verwendet, was dem Fiat 179 abging. (Photo: Daimler.)
198. 1972 sieht es im Mercedes ESF13 innen so aus. Die Knautschzone vorne und hinten fällt sofort auf. (Photo: Daimler.)

194. VW is able to infuse a little elegance into the 1971 ESVW, which would certainly have benefited the series. (Photo: Volkswagen AG.)
195. The 1981 Mercedes made little effort to combine safety and streamlining. (Photo: Daimler.)
196. Data for the 1972 Mercedes ESF13.
197. In its 1972 ESF13 model, Mercedes sparingly used the vulcanized rubber that the Fiat 179 lacked. (Photo: Daimler.)
198. Here's what the 1972 Mercedes ESF13 looks like inside. There is a conspicuous deformation zone in front and in the rear. (Photo: Daimler.)

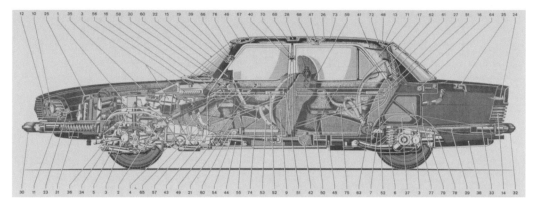

Technische Daten

Allgemeine Daten

Art des Aufbaus	Limousine
Zahl der Sitzplätze	5
größte Höhe unbelastet	1430 mm
größte Breite	1796 mm
größte Länge	5235 mm
Radstand	2850 mm
Sitzhöhe, unbelastet vorn	960 mm
Sitzhöhe im Fond	860 mm
Spurweite vorn	1444 mm
Spurweite hinten	1440 mm
Wendekreisdurchmesser	11,15 m
Kofferrauminhalt	0,51 m³
Fahrzeuggewicht fahrtig	ca. 2100 kg
Zulässiges Gesamtgewicht	ca. 2500 kg
Tankinhalt	ca. 80 l
Reifen, mit Schlauch	205 HR 14

AKTIVE SICHERHEIT
Fahrsicherheit

Motor
elektronische Benzineinspritzung
Abgasreinigungsanlage

Zahl der Zylinder	V6	
Gesamthubraum	2675 cm³	
Motorleistung nach DIN	140 PS	bei 5750 U/Min.
Motorleistung nach SAE	158 gr. HP	bei 5950 U/Min.
max. Drehmoment nach DIN	21 mkp	bei 4000 U/Min.
max. Drehmoment nach SAE	23 mkp	bei 4200 U/Min.

Lenkung
DB-Kugelumlauf-Servolenkung

Bremsen
Betriebsbremsanlage: Mercedes-Benz/Teldix-ABS-Syst.
fußbetätigte Feststellbremsanlage

Fahrwerk
Vorderachse: Doppelquerlenker mit Stabilisator
Hinterachse: Zweigelenk-Diagonal-Pendelachse mit Stabilisator und Niveauregulierung

Konditionssicherheit

Sitzkomfort
Gestaltung der Sitze unter Berücksichtigung der orthopädischen Erkenntnisse ; Steifigkeit und Dämpfung abgestimmt auf die Schwingungseigenschaften des Fahrzeugs; luftdurchlässige Sitzbezüge

Klimatisierung
Die Klimatisierung des Fahrzeugs wahrt die Kondition der Insassen durch:
ausreichende Sauerstoffzufuhr
Heizen oder Kühlen
optimale Temperaturverteilung
günstige Luftfeuchtigkeit

Schwingungs- und Geräuschverhalten
die Reduzierung der mechanischen und akustischen Schwingungen wird durch günstige Auslegung der Aggregate und der Karosserie, sowie durch geeignete Dämpfungsmaßnahmen erreicht

sign and safety, easily recognizable by rubber coverings on parts that are particularly vulnerable in accidents. The attached chart documents the extensive studies on this topic. The introduction of restraint systems such as the safety belt and airbag alone confronted the designers with new challenges in the car's interior. The spindly A, B and C pillar that supported the windows turned into thick posts made of high-tensile steel for greater crash stability. Additional newly evolving airbags in the sides and pillars limited the former wide panoramic view. The many examples of safety cars named here date back to 1972. To make it easier to understand we have not used the term »safety car«, which is used by Formula One when there is a crash and speeding cars have to be reined in. A second approach to increasing vehicle safety even more was the »Aktion Auto 2000« in 1981. The design of the VW contribution was a precursor of the Passat that followed it, which in spite of its front-mounted engine had no grill. There was the ESVW version of 1972, a mix between the VW Scirocco and the Ro 80, more appealing despite its thick rubber trim and hefty bumpers. Mercedes' Auto 2000 was a flat but clumsy version of the streamlined S-Class. In aerodynamic terms, the concept used a huge glass dome with a separation edge in the rear; the notchback sedan thus turned into a hatchback. But that's the thing about medicine: If it tastes good, people think it's not as good for them. It's an open question whether this is also true of many safety car concepts. Speaking of safety, we may say with certainty that the designers and naturally also the engineers, whose work tends to remain hidden from untrained eyes, transformed awkward safety cabins into cars that discreetly featured crash-decreasing components, which can be regarded as psychologically positive. For if one is constantly reminded by obvious fixtures how dangerous traffic is even today, one will tend to be far less enthusiastic about one's car. Here's food for thought: The risk of dying in a car is still a hundred times greater than the risk of dying in a plane crash.

Active and passive safety has been established as a standard, even now that we have many electronic aids to ensure safe driving conditions. A new process a driver must learn is how to correctly operate these aids. Navigation tools decrease the risk of unsafe orientation drives. New light systems with many variants and sensor-controlled exposure modes take into account oncoming traffic and improve driving at night. At the same time the designers are given new design opportunities, including overblown light graphics on the front and rear of the car. Operating technology becomes more exciting when, controlled by gestures, it has to transmit commands. Switches or buttons that were once responsible for injuries have become obsolete. However, control by means of gestures must be one hundred percent reliable, especially when it comes to driving commands. After all, how many panicky people react according to the rules with the prescribed gestures?

199. Der Chrysler Diablo kann 1957 die formale Ruhe sein, ohne langweilig zu wirken. (Photo: Wallpaperup.)
200. 2015 kann der Aston-Martin Taraf zeigen, wie man Lichter und Grill markentypisch vereint. (Photo: Aston Martin.)
201. Der Renault Espace macht es 2015 ebenso, übertreibt es aber etwas drei Preisklassen darunter. (Photo: Carricos.)
202. 2014 haut der Ford Edge die ganze Wucht des Bugs in den verchromten Grill. Big is beautiful, und das liebt man in den USA. (Photo: Ford.)

7. Form und Technik formen das Gesicht eines Autos

Der Wind war mitbeteiligt, als sich Scheinwerfer zwischen Kotflügel und Motorhaube vergruben. Das war in den 1930er Jahren, als die Aerodynamik lehrte, windschnittig sei nur die Form, die wenig störende Elemente in ihrer Blechhaut hat. Hilfreich und förderlich zeigte sich die Pontonform, die sich zunächst in sportlichen Versionen flach und flächenbündig bewährte. Mag sein, daß 1947 der Fiat 1100 mit Pininfarina-Karosse erste Ansätze der Verschmelzung von Lichtern und Kühlermaske hatte. Kreisrunde Leuchten auf identischer Höhe mit horizontalen Chromstegen, die die übliche Kühlermaske ersetzten. Kühlermasken gab es nur wenige wie bei Mercedes, die diese Maske formal ausbildeten wie den funktionsfähigen Kühler. Viele orientierten sich an Vorbildern aus den 1930er Jahren. Aerodynamisch geprägtes Nachdenken nahm Chromstreifen, belegte gerundete Fronten zu Lasten klassischer Kühlermotive. Beim Fiat fehlte die konsequente Verbindung, denn die Chromstege hielten noch Abstand zu den Leuchtengläsern. Die silbernen Stege beulten mittig etwas aus, als müßte der vergessene Kühler ein formales Echo erhalten. Konsequent hatte der Concept-Car Chrysler Diablo 1957 das neue Designmotiv durchgearbeitet. Ein starkes, umlaufendes Chromband öffnete sich im Bug zum schlanken, horizontalen Oval, darin lagen die Doppelscheinwerfer und ein Grill mit klaren senkrechten und horizontalen Stegen – ein Motiv, das bis heute gültig ist, weil die Basis klare geometrische Details aufweist. Opel hatte seinen Rekord, seine Kapitäne, Admiräle und Diplomaten 1964 ähnlich stringent gestaltet. Die Autos hatten glattflächige, auf halber Höhe mit einer Lichtkante differenzierte Flanken. Dieser Flächenknick war im Buggrill übernommen worden, beim Rekord mit runden Scheinwerfern, bei den großen Modellen mit Rechteckscheinwerfern. Die konsequente Formensprache wurde durchgehalten. Bald zeigte weltweit fast jedes Modell diese Designsprache. Ebenso wie der Ford Taunus 1969, der den Grill mittig vorschob, auch »Knudsen-Nase« genannt, wie seine US-Vorbilder. Namensgeber war der Ford-Chef Knudsen, er setzte dieses Design-Merkmal als Ford-typisches Gesicht durch. Egal, ob Fiat 124 oder Toyota Corona, Renault R4 oder Audi 100, Ro 80 oder russischer ZIL 114 oder, oder, oder – nur kein Porsche oder Rolls-Royce.

Diese Mode lief aus, weil sie wenig ausbaufähig war. Erst als das Gesicht stärker als markenprägend entdeckt wurde, übten Designer neue Motive, inspiriert von der Vielfalt der 1950er Jahre. Jede Marke wollte unverwechselbar werden. Das ist verständlich, solange das Ergebnis der Absicht entspricht. Wie so oft ist menschliches Verhalten von Vorbildern geprägt. Wenn in der Welt automobiler Gestalter ein neuer formaler Weg für Karosseriemotive auftaucht, bemühen sich andere, die auf den Leim des reizvollen Designs gekrochen sind, auf eigene Art zu reagieren. Sind es wenige, so weit so gut. Sind es viele, dann setzt sich der Begriff Mode durch. Am Beispiel der Leuchten-Grill-Kombination ist dies bestens zu belegen, weil in nur kurzen Zeitabschnitten viele, allzu viele gern der Mode folgen. Allein die letzten Jahre sind Beispiele dafür. Der Ford Edge, ein wuchtiger SUV von 2014 mit US-Wurzeln, verklammert Lichter und Grill mit armdicken Chrombalken, den Burschen für das Gelände mimend. Die Umfassungslinie der Leuchten-Grill-Kombination wird oben gerade durchgezogen und fällt schräg verlaufend zwischen den Scheinwerfern nach unten ab, im mittleren Bereich horizontal weitergeführt. Dieses Prinzip der Umfassungslinie lebt sich in feinster Linienführung bei edlen und teuren Marken aus. Für diese Gruppe ist der Aston Martin ein Beispiel. Sein Modell »Taraf«, welch eigenartiger Name, übernahm 2015 A-M-typische Elemente früherer Grillformen in zarter Andeutung.

Mit Schlitzaugen wirkt der Grill wuchtiger, seine Kraft mildern schlanke, horizontale Stege. Gar nicht milde zeichnete Honda seinen Sportler NSX von 2015. Hier konkurriert der schwarz lackierte Steg, einer hängenden Unterlippe ähnlich und als Wort zulässig, wenn vom Gesicht des Autos die Rede ist, mit wabenförmigem Grillgitter, das sich, kleiner geformt, in den monströsen Nüstern

7. Form and technology shape a car's look

The wind played a role when headlights were buried between the fender and the hood. That was in the 1930s, when aerodynamics taught us that only a form that has few impeding elements in its sheet-metal skin is streamlined. The pontoon shape proved to be helpful and advantageous, initially in sporty versions, flat and flush-fitted. It is possible that in 1947 the Fiat 1100, with a Pininfarina body, was the first to conflate the headlights and the radiator grill – circular lights at the identical height as horizontal chrome bars in lieu of the conventional radiator grille. There were few radiator masks like those designed by Mercedes, who formally developed this grille, as they also did a functional radiator. Many modeled their designs on 1930s cars. Due to aerodynamic considerations, designers took chrome strips and overlaid rounded front sections at the expense of classic radiator designs. In the Fiat, there was no logical connection, for there was still a distance between the chrome bars and the lamp lenses. The silver-colored bars bulged a little in the center, as though the forgotten radiator had to have a formal echo. The 1957 concept car Chrysler Diablo had consistently worked out the new design motif. A thick, circumferential chrome band opened in the front section into a slim, horizontal oval; in it were double headlights and a grille with distinct vertical and horizontal bars – a feature that is still valid today, because the base shows clear geometric details. Opel had designed its Rekord, its Kapitän, Admiral and Diplomat models just as stringently in 1964. The cars had slipstream flanks, differentiated halfway up by a light edge. This sharp break was extended to the front grille; the Rekord had round headlights, while the large models had rectangular ones. The consistent stylistic idiom was sustained throughout. Soon almost every model all over the world had the same styling – similar to the 1969 Ford Taunus, whose grille stuck out in the center, also dubbed a »Knudsen nose«, like its U.S. models. The name came from Ford's president Knudsen, who pushed through this design feature as a typical Ford look. Regardless of whether it was a Fiat 124 or Toyota Corona, Renault R4 or Audi 100, Ro 80 or Russian ZIL 114, every one of the models had this look – except for Porsche and Rolls-Royce.

This fashion petered out because it could not be developed further. It was only when designers more fully realized that the specific look determined brand recognition that they came up with new motifs, inspired by the wide variety prevalent in the 1950s. Every brand wanted to be distinctive. That's understandable as long as the result corresponds to the intention. As so often happens, human behavior is influenced by archetypes. Whenever in the world of auto designers a new formal way of styling bodywork appears, others who have been seduced by the charming design endeavor to interpret it in their own typical way. A few designers hardly create a trend. If many adopt a new style, the new fashion catches on. The best example of this is the combination of headlights and grille, because over relatively short periods many, all too many, have followed the fashion. There have been many examples of this in recent years alone. The Ford Edge, a hefty

199. The 1957 Chrysler Diablo can be formally sedate without appearing boring. (Photo: Wallpaperup.)
200. In 2015 the Aston-Martin Taraf shows how lights and grille can be combined in a way that is typical of the brand. (Photo: Aston Martin.)
201. So does the 2015 Renault Espace, though it exaggerates the feature at about three price brackets below the Taraf. (Photo: Carricos.)
202. The chrome grille in the front section of the 2014 Ford Edge creates a powerful impact. Big is beautiful – and that's what U.S. drivers love. (Photo: Ford.)

unterhalb der Schlitzaugen wiederfindet. Alle Details stammen aus dem Formenarsenal für Supersportwagen. Der Opel Astra 2015 spielt dagegen sein gestalterisches Repertoire korrekterweise nur vorsichtig aus. Bei gleichem Konzept der Umrißlinie spielen Scheinwerfer, Doppelchromstege und unterschiedliche Grillgitter behutsam das Thema Markengesicht durch.

Oder das Motiv reduziert sich stringent auf das Wesentliche. Bot der Renault Espace von 2015 noch ein Van-Konzept, das das ursprünglich vorbildliche aus den 1980er Jahren negiert, so haben die Designer für alle Renaults ein unverändertes Gesicht gezeichnet. Hier wird die Umrißlinie abgewandelt. Oben durchbricht der Rhombus die Horizontale, dafür wird unter den Scheinwerfern die Grillfläche knapper gehalten, ohne dem Rhombus zu nahe zu kommen. Kleine Renaults lassen den Grill zu einem schmalen Schlitz zusammenschnurren. Der Fiat Tipo bleibt dagegen 2016 nur halbherzig dem Thema der Leuchten-Grill-Kombination treu. Die untere Partie findet nur unbeholfen zu den Scheinwerfern, dafür glänzt der Grill mit frei stehenden, rechteckig gehaltenen Stegelementen, die vor Jahren bei Mercedes als punktförmige Zäpfchen zum »Sternchengrill« mutiert sind. Wie man mit Fiat-Grill und dem Motivthema danebengeraten kann, macht uns der Peugeot 5008 von 2017 vor. Zu viele Gestaltungsideen zerlegen die Grundform, so als hätten mehrere Köpfe um die Gunst der Gesichtsprägung gekämpft. Auch der kleinere 3008 macht es kaum besser.

Dankbar sollte man dagegen für das Gesicht eines 2016er Serien-Skoda Kodiaq sein, dessen Bruder von 2017 fast identisch ist, oder eines Hyundai Ioniq von 2016. Dieser fährt elektrisch und kann auf Gitter oder Stege ohne Gesichtsverlust verzichten. Bei den Studien hat die Limousine H600 von Pininfarina das Prinzip der Umrißlinien elegant umgesetzt, um diese mit senkrechten, leicht geknickten Stegen gleich wieder zu konterkarieren. Das wirkt wuchtig und findet sich bei Audi, BMW und Mercedes bei PS-starken Straßenfegern. Auch Mercedes kann sich diesem Thema nicht entziehen und gerät 2016 so ins Modische bei seinem SUV »Generation EQ«. Hier ersetzt eine Glasfläche mit liniengeprägter Lichtgraphik den klassischen Grill, was 2016 eher dem VW Budd-e zugestanden werden darf, der dezenter in der Bugschürze Lichtlinien einsetzt. Man mag sich nicht vorstellen, wie es wäre, wenn das serienreif gebaut würde. Wir bewegten uns so nächtens im Jahrmarkt der Lichteffekte, die ursprünglich dafür gedacht waren, eigene Marken im Dunkeln ausmachen zu können.

203. 2015 kann es der Opel Astra. (Photo: Opel.)
204. 2016: Fiat Tipo kann es. (Photo: Fiat.)
205. 2016: Pininfarina H600 kann es. (Photo: Pininfarina.)
206. 2016 kann es sogar die Generation Mercedes EQ. (Photo: Daimler.)
207. 2016 hält die Skoda-Studie Kodiaq mit. (Photo: Skoda.)
208. 2016 macht der VW Budd-e das Monster. (Photo: Volkswagen AG.)
209. 2015 grinst Honda NSX unbotmäßig. (Photo: Honda.)

203. The 2015 Opel Astra. (Photo: Opel.)
204. The 2016 Fiat Tipo. (Photo: Fiat.)
205. The 2016 Pininfarina H600. (Photo: Pininfarina.)
206. The 2016 Mercedes EQ can do it too. (Photo: Daimler.)
207. The 2016 Skoda Design Study Kodiaq keeps up the idea. (Photo: Skoda.)
208. The 2016 VW Budd-e makes like a monster. (Photo: Volkswagen AG.)
209. The 2015 Honda NSX sports a rebellious grin. (Photo: Honda.)

SUV built in 2014 that originated in the U.S., links lights and grille by means of chrome bars as thick as an arm, mimicking an all-terrain muscle car. The circumferential line of the lamp and grille combination is straight and continuous at the top and slopes downward transversally between the headlights to continue horizontally in the central section. This principle of the circumferential line is featured by top-of-the-range expensive marques. The Aston Martin exemplifies this group. In 2015 its model »Taraf« – an unusual name – adopted elements of former grille forms typical of Aston Martin, delicately hinted at.

With horizontal light clusters the grille looks more massive, its impact made milder by slim, horizontal bars. The design of the 2015 Honda NSX sports car was not mild at all. Here, the black lacquered ridge, resembling a hanging lower lip, and permissible as a word when spoken of the face of the car, competes with a honeycomb shaped grille, which is smaller in shape, in the monstrous nostrils below the slotted eyes. All details come from the array of forms used in styling super sports cars. More conventional, the 2015 Opel Astra on the other hand only cautiously shows off its repertoire of designs. In a design that uses the same circumferential line, the headlights, double chrome bars, and different grille lattices gently play the theme brand face.

Or else the styling is strictly reduced to essentials. While the 2015 Renault Espace still offered a van concept that negates the once exemplary design of the 1980s, the designers have created an unaltered look for all Renaults. Here the contour is modified. Above, the diamond shape breaks through the horizontal line; in return, the grille surface under the headlights is kept more skimpy, without getting too close to the diamond. In small Renaults, the grille contracts into a narrow slit. The 2016 Fiat Tipo only makes a half-hearted attempt at combining the headlights and the grille. The lower part only clumsily finds its way to the headlights, but the grille gleams, having free-standing rectangular bars that years ago, in Mercedes vehicles, mutated into a »star grille« composed of tiny cone-shaped dots. The 2017 Peugeot 5008 demonstrates how one can go wrong with a Fiat grille and spoil the image. Too many design concepts fragment the basic form, as if several people had struggled for the privilege of shaping the look of the car. The smaller 3008 model hardly did any better.

On the other hand, one should be grateful for the distinctive look of the 2016 production Skoda Kodiaq, whose 2017 brother is almost identical, or of a 2016 Hyundai Ioniq. The latter is electric and can get along without lattices or bars without losing face. In the studies, the H600 sedan by Pininfarina elegantly puts into practice the principle of the contour lines, only to counteract them at once with vertical, slightly buckled bars. The effect is powerful, and can be seen in the high-hp speedsters of Audi, BMW and Mercedes. Mercedes too cannot avoid following the trend, the result being its fashionable 2016 SUV »Generation EQ«. Here a glass panel with lines of light graphics replaces the classic grill, which in 2016 might be expected of a VW BUDD-e, though the latter more discreetly puts its light lines in the fairing. It is hard to imagine what would happen if this car was mass-produced. If it was, we would move about in a carnival of nocturnal light effects that were originally intended to allow people to tell one brand from the other in the dark.

7.1. Die Form wird kopiert, die Technik nicht

Dieses Kapitel beschränkt sich auf ein Bugdetail: ein Detail von Audi. Es fing vor vielen Jahren harmlos an. Da hatten Audis oberhalb der Prallfläche einen braven, spannungslosen Grill und im unteren Bereich der Frontschürze eine ebenso harmlose Luftöffnung. Ihre Form begann sich in der Breite dem Grill darüber anzupassen. Augen, die Zusammenhänge erkennen können, verbanden gedanklich oben und unten. Handwerklich Begabte haben mit dieser Erkenntnis ihren alten Audi auf neu getrimmt. Sie verbanden mit dünnen Chromstreifen oben und unten, lackierten die Fläche dazwischen schwarz und zauberten so bei flüchtigem Hinschauen einen neuen Audi, der diese Designikone des sogenannten »Singleframes« ab Werk lieferte. Mit dem Original gemein waren die immer noch getrennten Grillpartien, denn im mittleren Bereich war die verbliebene glatte Fläche nur zaghafte Prallfläche dank fehlendem Grillgitter und Träger für das amtliche Kennzeichen. Diese Frontgestaltung hatten so gut wie alle Audis vom A3 bis zum A8. Bleiben wir bei den Augen, dann konnte das ungeübte schwer ausmachen, wer denn da im Rückspiegel erschien, ähnlich wie bei den Familien von Mercedes und BMW. Ein Problem, wenn jedes Modell aus einem Haus sofort erkennbar sein sollte! Vermutlich erkannte Audi dies und rüstete etappenweise bei der Neueinführung von Modellen die Grillpartien um. War zum Beispiel beim Audi A6 von 2011 noch der erste Singleframe mit minimal abgeschrägten Ecken unten und oben etwas ausgeprägter, so war das Gesamtbild des Bugs schon recht wuchtig – so betörend betont, daß kopierwillige Designer das Motiv gerne aufnahmen. Besonders wichtig gerierten sich bei der chinesischen Marke der MG Roewe 750 von 2007 und seine kleineren Geschwister 550 und 459 mit von Audi inspiriertem Grill. Von den Chinesen und Japanern wissen wir, daß sie gerne kopieren, was selbst in unseren Breitengraden vorkommt (man lese und betrachte dieses Buch aufmerksam). Für die asiatischen Hersteller gilt es als Ehre, wenn man ein Vorbild zum Kopieren hat, sie sehen darin die Anerkennung einer fremden Leistung.

Zur Anlehnung an die großen Mäuler, die die neuen Mode prägten, kann man den Mitsubishi ASX von 2010 zählen, der den Ingolstädter Vorgaben nicht so vertraut, mit leicht nach unten verbreitetem Grill, nach vorne geneigt wie eine schnüffelnde Nase. Beibehalten wurde nur die mittlere Prallfläche zur Aufnahme des Kennzeichens. Das Nachfolgemodell hat sich aber von der dem Audi gleichenden Front verabschiedet und paßt sich den großen Brüdern an, die kräftige, horizontal gelegte Chrombügel aus den Seiten in die Front ziehen lassen, ein Motiv, das chromlos die Volvos seit 2016 zeigen. Audi merkte inzwischen, daß der Singleframe in seiner ersten Version zwar groß, aber wenig dynamisch erschien. Da hilft das Gestaltungshausmittel der schnittigen trapezoiden Formen: Die winzigen, abgeschrägten Ecken wachsen zu neuer Größe, und der Singleframe gerät so zur Kombination zweier Trapeze, das obere kleine verjüngt sich aufwärts, das größere als Basis geht nach oben in die Breite. Sehr gut, auch ohne Worte dargestellt, ist das

210. Der Chinese Voleex C70 der Marke Great Wall ist 2011 noch ruhig. (Photo: Great Wall.)
211. Der Audi A6 beginnt 2011 das Spiel mit dem Grill. (Photo: Audi.)
212. Das Coupé Haval von Great Wall wird 2015 lauter. (Photo: Great Wall.)
213. Die Studie Viziv von Subaru macht 2013 den Trend mit. (Photo: Subaru.)
214. Der Audi Q5-Grill läßt 2016 die Scheinwerfer an sich heran. (Photo: Audi.)

210. The Chinese Voleex C70 (Great Wall Brand) – sedate and easygoing in 2011. (Photo: Great Wall.)
211. The 2011 Audi A6 experimenting with the grille. (Photo: Audi.)
212. Great Wall's 2015 Coupé Haval – a lot less sedate. (Photo: Great Wall.)
213. The 2013 Subaru study Viziv joins the trend. (Photo: Subaru.)
214. On the 2016 Audi Q5, the grille is right next to the headlights. (Photo: Audi.)

7.1. The form is copied, though not the technology

This chapter is limited to one front-section detail: a detail of the Audi. The whole thing started harmlessly many years ago. At the time Audis had a plain tensionless grille above the bumper system and an equally harmless air vent in the lower part of the lower front panel. Their form began to adapt to the width of the grille above it. Expert eyes made a connection between the upper and lower part. With this in mind, dexterous people put new trim on their old Audis. They connected the upper and lower part with thin chrome strips, painted the surface between them black and thus conjured up what looked like a new Audi to the casual observer. Audi delivered these design icons – the so-called »Singleframe grille« – directly from the factory. As in the original, the parts of the grille were still separate, for in the central section the remaining smooth surface was only a tentative deflector surface thanks to the missing grille lattice, and bore the official vehicle registration number. Practically all Audis, from the A3 to the A8, had this front design. Meanwhile, untrained eyes could hardly make out what type of car appeared in the rearview mirror; the same was true of the Mercedes and BMW families. It was a problem if a company's every model had to be immediately recognizable! Presumably Audi realized this and gradually retrofitted the grille sections when new models were introduced. For instance, while in the 2016 Audi A6 the first Singleframe, with minimally beveled corners below and above, was still somewhat more distinctive, the overall appearance of the front was already quite massive – so bewitchingly emphatic that designers who were eager to copy it were only too happy to adopt the motif. Prominent among these were the 2007 MG Roewe 750 from China and its smaller siblings 550 and 459, with a grille inspired by Audi. We know the Chinese and Japanese like to copy, which even happens in our latitudes (please read and examine this book carefully). Asian manufacturers consider it an honor to copy a model, regarding this as the recognition of someone else's achievement.

Among the imitators of the bigmouth models typical of the new fashion trend is the 2010 Mitsubishi ASX; not entirely familiar with the specifications of the Ingolstadt manufacturer, the ASX had a grille that became slightly wider toward the bottom, and tilted forward like a sniffing nose. It retained only the center deflector plate where the registration number was displayed. The front of its follow-up model, however, no longer copies the Audi, choosing to conform to its big brothers, which have hefty, horizontal chrome-plated bars extending into the front from the sides, a feature that the Volvos have been sporting since 2016. In the meantime Audi has realized that while the Singleframe in its first version appeared big, it did not look very dynamic. This calls for the designers' household remedy – sleek trapezoid shapes: The tiny beveled corners are given a new dimension, and the Singleframe thus becomes a combination of two trapezes, with the small one at the top tapering upwards, while the bigger one, the base, gets wider toward the top. The one

im Bild des Audi Q5 von 2016, einem SUV-Ableger der Audi-A6-Limousinen. Die Modelle Voleex C70 von 2011 und Haval Coupé C von 2015 machten aus der geradlinigen Vorgabe minimal geschwungene Umrisse und vernebelten den Audi-Effekt mit kräftigen horizontalen Chromstegen, eher passend für einen SUV.

Wenn Concept-Cars Motive anderer übernehmen, so ist das die Erkenntnis, wie gut Vorbilder für Automobile sind, die sich modern, dynamisch und exklusiv geben wollen. Designer greifen dankbar danach, variieren etwas und sehen sich so vermeintlich auf der Höhe der Zeit. Die Studie Subaru Viziv von 2013 macht das behutsam und beläßt den Grill in typisch trapezförmigem Umriß und zurückhaltender Größe, um den hakenförmigen Lichtschlitzen Platz zu lassen. Die wiederum sind gern verwendetes Motiv, das mal elegant wie beim Cadillac Escala 2016 zu sehen ist oder etwas kleiner beim Renault Megane 2016 und dem Honda Clarity Fuel Cell 2017. Der Erkenntnis wirkmächtig dynamisierter Trapeze folgt konsequent die Marke Hyundai. Der Chefdesigner des Kia-Hyundai-Konzerns dankte 2006 bei VW höflich ab, als seine Talente falsch eingeschätzt wurden, was im nachhinein viele Wolfsburger Bosse bedauerten. Seither pendelt Peter Schreyer, vor VW noch Designchef bei Audi, nun zwischen Frankfurt, der deutschen Kia-Zentrale, und Korea hin und her. Seine Arbeiten lassen sich in ständig beherrschteren Formen beider Marken ablesen, die auf dem besten Wege zu internationalem Standard sind. Das bisherige Meisterstück ist wohl die Oberklasselimousinenstudie Hyundai Vision G von 2016. Und in der Serie dürfen die i10 seit 2013, die i20 seit 2014, die i30 und i40 seit 2015 dieses Motiv in die Welt tragen. Bei so viel von Audi inspirierten Gesichtern ist es schade, daß Hyundai sich damit gleichmacht mit anderen Kopisten.

Da ist es belanglos, wenn sich ein Subaru Serien-Viziv von 2016 in der Audi-Grillkopie einen kräftigen Chromsteg gönnt, um das Firmenlogo aufzunehmen. Erst dann werden solche ansatzweise hilflosen Versuche zur eigenständigen Wirkung kommen, wenn alle Modelle einer Marke sich Inspirationen gönnen, diese aber so eindrucksvoll gestalten, daß es ihnen gelingt, ihrer Marke ein dauerhaftes Image der Gestaltung und Formensprache zu geben. Voraussetzung ist die Kraft, eine gefundene Lösung langfristig einzuhalten, statt zu vielen Trends zu folgen. Selbst wenn sie mit der Gefahr leben, über alle Modellreihen hinweg wenig zu differenzieren und damit Unterschiede zu verwischen, machen Mercedes, BMW und Audi klar, daß das Festalten an Designregeln mit behutsamen Modernisierungen nicht der falsche Weg ist. Ganz falsch lag Suzuki 2017 mit seinem kleinen Swift nicht. Er nutzte das Audi-Motiv für einen kompakten Grill, dessen harmloser Luxus die feinen, unterschiedlich geformten Grillstäbe sind, veränderte sonst aber die Vorgänger-Karosserie kaum. Ganz klein macht sich das Schnäuzchen des Datsun Go Cross 2016, ohne den Vorbild-Grill zu verändern, der aber ein SUV-ähnliches Gesicht zeigt. Stark verändert hat sich die Designsprache im Hause Citroën. Hier versucht die Marke seit langem, die einstigen ingeniösen und formalen Talentbeweise in neue Designsprache umzusetzen. Das fing mit dem DS3 an, dann bot uns der Cactus Flankenpolster. Extreme Studien unterstrichen den Willen für

215. 2015 sagt sich der Hyundai Vision G: Was Audi kann, kann ich (fast) besser. (Photo: Hyundai.)
216. 2017 will auch der kleine Suzuki Swift den Trend des Audi-Grills nutzen. (Photo: Suzuki.)
217. 2016 kann die Studie Datsun Go Cross dem Grillfest nicht widerstehen. (Photo: Datsun.)
218. Der Citroën DS7 Crossback vergißt 2017 französischen Chic und grillt mit. (Photo: Citroën.)
219. Der Viziv von Subaru trägt 2016 mit der Variante einer Chromspange dick auf. (Photo: Subaru.)

215. The 2015 Hyundai Vision G: Anything Audi can do, I can do better – well, almost. (Photo: Hyundai.)
216. Even the little 2017 Suzuki Swift gets in on the trend of the Audi grille. (Photo: Suzuki.)
217. The 2016 Datsun study Go Cross can't resist the grille craze. (Photo: Datsun.)
218. The 2017 Citroën DS7 Crossback forgets about French elegance and goes for the grille. (Photo: Citroën.)
219. The 2016 Subaru Viziv tries to impress customers with its version of a chrome bar. (Photo: Subaru.)

shown in the picture of the 2016 Audi Q5, an SUV spin-off of the Audi A6 sedans, is very good, even when not accompanied by text. In the Voleex C70 (2011) and Haval Coupé C (2015) minimally curved contours replaced the straight lines of their Audi model and obscured the Audi effect with hefty horizontal chrome bars, which would have been more suitable for an SUV.

When concept cars borrow the motifs of other marques, it is because designers realize how good it is to have models for modern, dynamic and exclusive-looking cars. Designers gratefully grab at novel features, modify them a little and feel they've reached the state of the art. The 2013 study Subaru Viziv does so cautiously, keeping the grille typically trapezoid and modest-sized in order to leave room for the hook-shaped light slits. The latter are a favorite motif – elegant, as in the 2016 Cadillac Escala, or somewhat smaller in the 2016 Renault Megane and the 2017 Honda Clarity fuel-cell car. The Hyundai brand is consistently based on the realization of the importance of powerfully dynamic trapezoids. In 2006 the head designer of the Kia/Hyundai company politely resigned from VW when his talents were misjudged, a fact that was later regretted by many Wolfsburg CEOs. Ever since, Peter Schreyer, who was head designer at Audi before he came to VW, has been shuttling back and forth between Frankfurt, the German headquarters of Kia, and Korea. His work can be recognized in the ever more restrained forms of both brands, which are well on their way to becoming an international standard. His masterpiece up to date is probably the 2016 luxury-class sedan study Hyundai Vision G. And the mass-produced i10 (2013), i20 (2014), and i30 and i40 (2015) carry this motif into the world. When the look of so many cars has been inspired by Audi, it is a shame that Hyundai was also modeled on it and thus chooses to be no different from other copyists.

At that point it's irrelevant if a 2016 Subaru production Viziv incorporates a hefty chrome bar, a copy of the Audi grille, for the purpose of displaying the company logo. Such initially helpless attempts will not effectively give the model a distinctive look until all the models of a brand express their designer's inspiration so impressively that their brand comes to have a lasting design and a style of its own. This means that designers have to have the tenacity to stick with a solution, once found, over the long term instead of following too many trends. Even if they run the risk of differentiating all the series of models only slightly and thus blurring distinctions, Mercedes, BMW and

neue Formen. Jetzt kann Citroën einen SUV als DS7 anbieten, der sich 2017 nichts daraus machte, einen von Audi inspirierten Bug zu zeigen, welcher auch im vorangegangenen Kapitel hätte stehen können.

7.2. Die Technik schenkt der Form die Flanken

Es geschieht selten, daß etwas im Automobildesign vergessen wird oder zu wertlos erscheint, um beschrieben zu werden. Jahrelang blieb ein Bilddokument vergessen, das ein wichtiges erstes Beispiel für bewußt gestaltete Flanken des Automobils ist. Die Berliner Automarke Szawe, Szabo & Wechselmann, mit hochpreisigen Modellen, lebte nur von 1920 bis 1924, die Zeit nutzend für eine neue Idee. Viele Autos hatten bereits die Torpedoform und bauähnliche Karossen, beides mit glattflächigen Flanken. Ernst Neumann-Neander, der Vater der »Fahrmaschine«, hatte als Graphiker und Gestalter jenes Feingefühl, um bei der mathematisch inspirierten, einer Parabel ähnlichen Form Qualitäten zu erkennen. Es lag nahe, dieses Motiv am Auto umzusetzen, die Möglichkeit dafür bot Szawe. Das präzise Jahr ist unbekannt, vermutlich war es 1920 oder 1921, als das erste Modell mit dieser Flankengestaltung auftauchte: die liegende Parabel, nach hinten spitz auslaufend, was optische Dynamik generierte. Gekoppelt mit Scheibenrädern, wirkte das dachlose Auto glattflächig elegant. Der Szawe 1038 von 1921 hatte das Parabel-Motiv umgekehrt vorne angespitzt. Erst Jahrzehnte später haben GM-Designer diese optische Wirkung wiederentdeckt und schmückten Corvette-Flanken damit, hervorgehoben durch zweifarbige Lackierung. Der Szawe Type 109 begnügte sich mit leichter Verfaltung der Flankenbleche und verlor so die Dynamik, paßte aber in die Art-Déco-Zeit. Ein anderer Berliner, Josef Winsch, hatte 1921 mit seinem Modell Joswin June die Szawe-Parabel übernommen. Unbekannt bleibt, ob auch hier Neumann-Neander seinen Stift führte.

Einen großen Sprung machte die Gestaltungsaufgabe der Flanken. Noch waren üblicherweise die Karosserien, offen oder geschlossen, einander ähnlich mit wenig inspirierenden Bauteilen, um die Seiten besonders zu formen. Es langten Trittbretter, darauf verstaute Werkzeugkästen, Kotflügel und Türen, um die Seitenansicht zu differenzieren. Mit Aufkommen der selbsttragenden Karosserie und der aerodynamisch beeinflußten Form gab es glatte Seitenpartien. Dicke Blechstärken brauchten zur Flächenstabilisierung noch keine Falten oder Sicken. Das änderte die Einführung der Pontonform, die anfänglich ungestörte, glatte Flanken hatte. Man kann trefflich streiten, ob der Horror vacui, die Angst vor der Leere, der Ansporn war, diese glatten Seiten mit blechgeprägten Linien zu beleben. Dies bleibt offen, und das sehr bekannte Beispiel des Studebaker Starliner von 1953 darf gelobt werden ob der Idee, eine Linie von vorne bis vor das Hinterrad zu zeichnen, wo sie im formalen Echo der hinteren Seitenfenster noch im Türbereich nach unten zieht. Selten ist seither eine so geglückte, dauerhaft in Blech geformte Linie gezeichnet worden. Und elegant war der Studebaker ohnehin. Raymond Loewy als Designer dieses Autos vermied die wuchtige Erscheinung der Konkurrenzmodelle und wurde wegen der italienischen Leichtigkeit gelobt: flache Haube, schlanke Lufteinlässe, später abgewandelt bei den Alfas zu sehen, und Panorama-Heckfenster, das bei uns erst 1957 den Opel Rekord auszeichnete.

Flanken können das Parabelmotiv abändern, wie beim Ferrari 375 MM von 1954. Hier wird nicht die Fläche ausgeformt, hier liegt auf glattem Grund eine parabelähnliche Sicke, die aber we-

220. Kein Amerikaner ist 1953 zurückhaltend vornehmer als der Studebaker Starliner. Das verwundert bei seinem Designer Raymond Loewy, der sonst nicht abgeneigt war für zarte, aber verkraftbare Übertreibungen. Bei ihm vereinen sich Anfang und Ende: geboren in Paris, gelebt in den USA, gestorben in Monte Carlo. (Photo: Studebaker.)
221. Neumann-Neander zeigt 1924 seine Idee der Flankengestaltung am Szawe 120/50 PS. (Photo: Szawe.)
222. Der Szawe 109 zeigt 1922 eine Variante mit Flankenfalten. Im Vergleich dazu ist die Parabelform ausgewogener. (Photo: Szawe.)
223. Der Joswin June kann 1921 dieselbe Idee aufweisen wie Neumann-Neander 1922. Frage: Hat N.N. das gesehen, oder setzte sich eine Idee unabhängig in den Gestalterköpfen fest? Wer weiß es? (Photo: Bryunzeel.)

Audi make it clear that adhering to design rules while cautiously modernizing the designs is not the wrong way forward. Suzuki, in 2017, was not all that wrong after all with its little Swift. The Swift used the Audi motif for a compact grille whose harmless luxury lies in the delicate, differently shaped bars of the grille, but hardly modified the bodywork of its predecessors. The little nose of the 2016 Datsun Go Cross is quite small, without altering the grille it is modeled on, but it has the look of an SUV. Citroën's design idiom has changed substantially. For a long time, the brand has been trying to translate the ingenious and formal talented ideas of the past into a new design language. It all began with the DS3; then the Cactus offered us side airbumps. Extreme studies emphasized a wish for new forms. Now Citroën is able to offer an SUV, the DS7, which in 2017 was not bothered by the fact that it had a front section inspired by Audi, one we could have included in the previous chapter.

7.2. Technology determines the form of the flanks

It rarely happens that something about a car's design is forgotten or seems too worthless to describe. For years a picture documenting an important first example of the intentionally designed flanks of a car was forgotten. The Berlin auto brand Szawe, Szabo & Wechselmann, which had high-priced models, was in existence only from 1920 until 1924, using the time for a new idea. Many cars already had the torpedo form and bodywork that was similar in design, with slipstream flanks. Ernst Neumann-Neander, the father of the »Fahrmaschine« (driving machine), was a graphic artist and designer, and was sensitive enough to recognize the qualities of the mathematically inspired, parabola-like form. It made sense to put this motif into practice on a car, and Szawe provided an opportunity for him to do so. The exact year when the first model with this flank design first appeared is unknown – presumably 1920 or 1921: the recumbent parabola extending backwards into a sharp point, which generated visual dynamics. Coupled with disc wheels, the roofless car looked streamlined and elegant. Conversely, the 1921 Szawe 1038 had the pointed parabola motif in front. It wasn't until decades later that GM designers rediscovered this optical effect and decorated the Corvette's flanks with it, emphasized by two-tone paintwork. The Szawe Type 109 had slightly folded metal flank panels and thus lost the dynamism, but was in step with the Art Deco period. Another Berliner, Josef Winsch, had adopted the Szawe parabola in 1921 with his model Joswin June. It is not known whether Neumann-Neander was also involved in designing it.

The design of the flanks made a great leap forward. At the time the bodies of the cars, open or closed, usually still resembled each other, with uninspiring components used to form their sides. Only running boards, toolboxes tucked away on them, fenders and doors differentiated the side profile. Once the self-supporting body and an aerodynamically influenced form first appeared, there were now smooth side panels. Thick metal panels did not need folds or swage lines to stabilize them. This made a difference to the newly introduced pontoon form, which initially had uninterrupted, smooth flanks. One could argue endlessly whether it was *horror vacui*, the fear of empty space, that was the incentive for adding lines stamped into the metal to these smooth sides. The question remains open, and we must commend the well-known example of the 1953 Studebaker Starliner because it had the idea of drawing a line from the front to just before the rear wheel where, formally echoing the back side windows, it moves downward while still in the door panel. Rarely since then has so successful a line been drawn, permanently stamped in metal. And the Studebaker was elegant in any case. Raymond Loewy, the designer of this car, was careful not to give it the bulky look of its competitors and was praised for its Italian lightness: a flat hood, slender air inlets, later found modified in the Alfas, and a wrap-around rear window that here in Germany did not appear until 1957 in the Opel Rekord.

Flanks can alter the parabola motif, as in the 1954 Ferrari 375 MM. Here it is not the surface that is molded: On a smooth foundation we have a parabola-like swage line that is less elegant, however. When using flanks in order to call up memories of older car bodies, designers cite chubby-cheeked fenders, particularly over the rear wheels. Like the Karmann Ghia, the 1955 Borgward Isabella, a coupé version, is an example of this frequently encountered design feature. Or the flanks are practically left alone, while the wheel cutouts are enlarged far more than necessary. This is most clearly apparent in the 1955 concept car Pontiac Strato Star. The color of the interior surfaces of the visible wheel well effectively contrasts with that of the car, as evidenced by many GM variants. Also painted a contrasting color was the parabolic surface of the 1957 Corvette, which used this design idea all over the world as a distinctive feature. How absurd the little

224. 1955 will die Studie Pontiac Stratostar die Balance finden zwischen leicht und schwer, rund und gestreckt, Chrom und Lack, aber mit eindeutig überbetontem vorderen Radausschnitt. (Photo: GM.)

225. Der Triumph TR8 vergißt 1980 seine englische Noblesse der konservativen Haltung und macht mit Keilform und Seitenschmiß in der Flanke auf neumodisch. (Photo: Brooklandscc.)

226. Das muß man 2005 können: eine Dachrinne mit Frontlicht und Flankenprägung kombinieren. Die Studie Suzuki Ionis probiert es. (Photo: Suzuki.)

227. Der Ferrari 375 MM kann 1954 leider nicht so elegant die Flanke prägen wie der Szawe von 1924. (Photo: Conceptcarz.)

228. Der brave Lloyd aus Bremen als Alexander will 1959 den eleganten Italiener im Kleid von Frua geben. Bei dem Radstand schwierig. (Photo: P. Weiser.)

229. Die Corvette von Chevrolet macht 1957 die Flankenparabel weltberühmt, wobei die Farbe mithalf. Und Szawe war bereits längst vergessen. (Photo: Wheelsage.)

230. Der Stanguellini Spider von Bertone kupfert 1957 die Corvette. (Photo: Zagatoff.)

niger elegant verläuft. Flanken zu nutzen, um an ältere Karosserien zu erinnern, nutzt das Zitat pausbäckiger Kotflügel, speziell über Hinterräder. Die Borgward Isabella als Coupéversion von 1955 ist wie der Karmann Ghia Beispiel für diese oft gesehene Gestaltung. Oder die Flanken bleiben so gut wie unangetastet. Dafür werden die Radausschnitte stärker als notwendig vergrößert. Zu erkennen ist dies am Concept-Car Pontiac Strato Star von 1955. Die sichtbar gewordenen Innenflächen des Radhauses sind effektvoll farbig abgesetzt von der Wagenfarbe, was GM-Modelle variantenreich belegten. Ebenso farbig abgesetzt wurde die Parabelfläche der Corvette 1957, die diese Designidee weltweit als Wiedererkennungsmerkmal nutzte. Wie albern wirkt da der kleine deutsche Brütsch Jet 2VN von 1958 im Kunststoffkleid mit Corvette-Motiv, einer Methode, preiswert und schnell eine Autohaut zu kreieren. Der Fiat 1200 Spider aus dem Corvette-Jahr, gezeichnet von Bertone, gebaut von Stanguelini, hatte eine ins Blech geprägte Parabel mit passabler Proportion und Verzicht auf farbige Betonung, was italienischer Eleganz geziemt. Italienisch, aber wegen kurzen Radstands des Originals weniger gelungen, war auch die Hülle des Lloyd Arabella 1959. Der Designer Pietro Frua, einer der ganz wenigen Soloarbeiter in diesem Metier, nutzte die Designmotive seiner Renault Caravelle aus demselben Jahr. Die Flanken hatten kräftig betonte, horizontale Sicken entlang der Wagenlänge, die, dicht übereinanderliegend, die Flächenstabilität förderten. Zu loben aber ist neben den Flanken die Kombination vom Grill mit innen hochgezogenen Stoßstangen. Das verchromte Motiv sahen wir beim Allzeitklassiker Lancia Aurelia B24 Spider von 1954.

Flanken mit geprägten Motiven sind formal gelungen oder auch nicht. Gute wie schlechte Beispiel haben einen Vorteil gemeinsam. Blechschäden lassen sich handwerklich bei sicken- oder faltenzerfurchten Flächen leichter beheben. Hier fallen minimale Unterschiede in der Oberfläche nicht so auf wie vergleichsweise bei einer glattflächigen, bombierten Fläche eines Porsche 356 A. Seine lackierten Flächen verraten jeden Blechpickel, jede schlampig ausgebeulte Delle. Bleiben wir bei einem Flankenbeispiel der zwiespältigen Art: Der Triumph TR8 von 1980 hatte eine dominante Keilform, was dynamisierend ausreichte. Er wollte nicht auf einen kräftigen Schmiß in der Seite verzichten, als säbelte der Designer beim Clay-Modell durch die Flanke. Ganz anders war die Seitenansicht eines Autos mit ungewohnter Tür. Der BMW Z1 von 1990 mit Kunststoffkarosse läßt die kleinen Türflächen auf der Innenseite der Flanken verschwinden und darf mit der Lücke in der Gürtellinie trotzdem fahren. Lückenlos nach der Jahrtausendwende entwickelte sich das Thema Flankengestaltung, egal ob zurückhaltend oder überbordend. 2005 zeigte sich der Concept-Car Suzuki Ionis mit einer aus der Dachlinie nach vorn fallenden Kurve, die sich auf halber Höhe im Richtungswechsel in der Flanke weitergräbt. Ähnlich verwegen gaben sich riesige Gläser der Frontscheinwerfer. Wer's mag! Diszipliniertes Design zeigte dagegen der Concept-Car Opel Flextreme 2007, eine Idee, die serienreif im Zafira übersetzt wurde. Opel begann hier, mit einem elegant gewinkeltem Schwung mit horizontalem Auslauf markeneigene Linien zu etablieren. Die Studie Hyundai HED4 vom selben Jahr nutzte die Freiheit, die ein Concept-Car hat: hier die Variante, die Glasflächen der Seitenfenster sichelartig in die Seiten rauschen zu lassen. Viel mehr Form als logisches Design!

Dies bot die E-Klasse von Mercedes 2009 insgesamt, nur vergaß sie ein klares Design oberhalb der Hinterräder. Hier zitierte man die heroischen Zeiten der 1950er Jahre, als bei den Mittelklassemodellen 190 bis 220 die Stilistiker schamhaft die Konturen von geschwungen Kotflügellinien nachzeichneten, die vor den Hinterrädern im Viertelkreis nach unten liefen. Die Mercedes-Werbung lobte den Blick zurück, praktisch aber ein albernes Motiv, was man nach sechs Jahren erkannte und mit minimalem Trick im hinteren Türblech die Linie geradzog. Geht doch. Das ging aber nicht, wenn die Coupé- und Cabrioversion nur viertürig war und die kleine Peinlichkeit behielt. Ganz abwegig wurde in der B-Klasse und in der neuen A-Klasse 2015 ein Flankenschwung

German 1958 Brütsch Jet 2VN looks clad in plastic with a Corvette motif, one method of cheaply and quickly creating a car's shell. The Fiat 1200 Spider, manufactured during the Corvette year, designed by Bertone and built by Stanguelini, had a parabola stamped into the steel, of decent proportion and without color accentuation, as befits Italian elegance. Also Italian, but less successful because of the original's short wheelbase, was the body of the 1959 Lloyd Arabella. The designer, Pietro Frua, one of the very few solo workers in this profession, used the design motifs of his Renault Caravelle, built the same year. The flanks had boldly emphasized, horizontal swages along the entire length of the car; lying directly above one another, these promoted surface stability. But what is commendable, beside the flanks, is the combination of the grille with retractable bumpers. The chrome motif is something we saw in Lancia's 1954 all-time classic, Aurelia B24 Spider.

Flanks with stamped motifs are formally successful, or not. Both good and bad examples have one advantage in common. Car body damage can be repaired more easily in surfaces that are furrowed by swages or folds, as minimal surface differences are not as conspicuous there as, say, in the smooth, convex surface of a Porsche 356 A. Its painted surfaces reveal every metal blister, every sloppily popped dent. Let's look at a flank of the ambiguous kind: The 1980 Triumph TR8 had a dominant wedge shape, which made it sufficiently dynamic. It did insist on a hefty gash in its side, as if the designer had slashed the flank of the clay model with a saber. The side view of a car with an unusual door, the 1990 BMW Z1, was totally different – it has a plastic body, and the small doors drop into the door sills, but, with a gap in its waistline, it is still allowed on the road. After the turn of the millennium flank design – conservative or exuberant – has been developing apace. In 2005 the concept car Suzuki Ignis sported a curve that falls forward from the roofline and, halfway up, changes direction to dig its way into the flank. Equally daring were its huge front headlights. Chacun à son goût! On the other hand, the design of the 2007 concept car Opel Flextreme was disciplined, an idea that was ready to go into production and was implemented in the Zafira. Here, Opel began to establish its proprietary lines with an elegantly angled sweep that tapers off horizontally. That same year, the study Hyundai HED4 took advantage of the freedom provided by a concept car: pictured here is the variant that has sickle-shaped, glass side windows. Far more form than logical design!

The 2009 E-Class Mercedes lacked clear design above the rear wheels. This section was an allusion to the heroic 1950s, when in the mid-range models from 190 to 220 the stylists bashfully traced the contours of curving fender lines that ran downward in a quarter circle in front of the rear wheels. Mercedes' advertising praised the retro look, but in practical terms it was a ludicrous

224. The 1955 Pontiac study Stratostar tries to find the balance between light and heavy, round and elongated, chrome and paintwork, but with a clearly overemphasized front wheel cutout. (Photo: GM.)

225. The 1980 Triumph TR8 forgets its aristocratic British conservative attitude and goes for a modern wedge shape, and a slash in the side. (Photo: Brooklandscc.)

226. Moving with the times: The 2005 Suzuki study model Ionis combines roof drip molding with front lights and side stamping. (Photo: Suzuki.)

227. Unfortunately the sides of the 1954 Ferrari 375 MM are not as elegantly stamped as those of the 1924 Szawe. (Photo: Conceptcarz.)

228. A good old 1954 Lloyd Alexander from Bremen trying to look like an elegant Italian model with a Frua body. Not easy with a wheelbase like that. (Photo: P. Weiser.)

229. The 1957 Chevrolet Corvette makes the parabola of the sides world-famous, and the color certainly helps. By then, the Szawe is a thing of the past. (Photo: Wheelsage.)

230. Bertone's 1957 Stanguellini Spider copies the Corvette. (Photo: Zagatoff.)

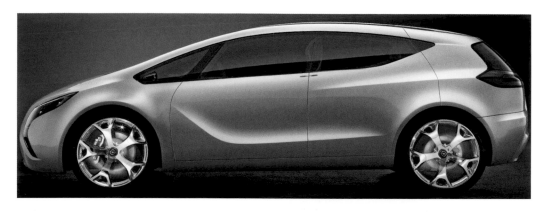

231. 2007 ist die Opel-Studie Flextreme zu loben. (Photo: Opel.)
232. 2015 fehlten beim Mercedes A AMG wohl die Designer vor dem Serienstart. (Photo: Daimler.)
233. Der Hyundai HED4 treibt es 2007 wild mit dem Glas. (Photo: Hyundai.)
234. Der Lada Vesta Cross Concept macht sich 2015 Mut nach Jahrzehnten der Öde. (Photo: Lada.)
235. 2017 macht der Renault Captur den Schweller zum Star der Flanke. (Photo: pinthiscar.)
236. Der Ferrari 812 Superfast kann 2017 die Flanke wuchtig wellen. Gerade noch akzeptabel. (Photo: Ferrari.)
237. Der Bentley GT Speed haucht 2016 Kotflügel-linien der guten alten Zeit in die Seite. Paßt. (Photo: Bentley.)
238. Das Kupferband der Sympathie fädelt sich 2017 durch den Body des Quant 48 Volt. (Photo: Quant.)

ins Blech getrieben, der zu nichts paßte, weder als Linienecho zur Dachlinie noch zu anderen Motiven. Er wurde aber von den Designern schöngeredet. Der Chef Gordon Wagener hatte für neue Modelle später den Begriff der »sinnlichen Klarheit« geprägt – nun ja, vielleicht passend für die neuen, aber kaum für die Unglücksraben zuvor. Und gern in den Pressetext aufgenommen, wird von der DNA einer Marke gesprochen, was bedeuten soll: nie die Herkunft ablehnen.

Man glaubt es nicht, aber auch Russen können, wenn sie wollen, vorausgesetzt sie heuern einen westlich geprägten Designer an, in diesem Fall Steve Mattin, einen Ex-Mercedes- und Ex-Volvo-Designer. Lada, am unteren Ende aller Preisskalen, hatte sich einen Concept-Car geschenkt, den 2015 niemand von dieser Marke erwartet hätte. Der Vesta Cross Concept hat zwar eine neue und ungewohnte Buggestaltung wie die neuen Mitsubishis mit gewelltem Grillgitter. Übermütig aber sind die Sicken, die über dem Vorder- und dem Hinterrad wie ein Bumerang mit einseitig kurzem Ende ins Blech gestanzt wurden. Gemessen an früheren Ladas, ist dies geradezu ein Jungbrunnen für das kaum veränderte Serienmodell. Am anderen Ende der Preisskala ist es immer gut, noble Zurückhaltung zu üben. Da verlegt ein Rolls-Royce eine Linie tief nach unten, die sich nur minimal hinter dem Vorderrad nach oben beugt. Und der frühere baugleiche Bruder und heute Teil des VW-Konzerns, der Bentley GT Speed, macht das, was einer E-Klasse nicht so gut gelang: das Nachzeichnen eines alten Bentley-Kotflügels aus den 1950er Jahren, was wir als Motiv schon öfter entdeckt und beschrieben haben. Jetzt schreiben wir das Jahr 2016, angemessen für ein der normalen Welt entrücktes Auto. Auch entrückt, aber 2017 nur eine Studie, ist der Quant 48Volt der Firma NanoFlowcell aus der Schweiz. Hier wird ein Auto gezeigt, das wie eine Fließheck-Luxuslimousine wirkt und gut sein soll für 300 km/h, und das mit rein elektrischem Antrieb. Dies wird unterstrichen durch ein Kupferband – aufgemerkt: Das ist der Hinweis auf die Kupferdrahtwicklung des E-Motors. Dieses Band, über dem Heckfenster gewellt und damit der Dachform folgend, schiebt sich in der Flanke hinter das Blech, um weiter unten wieder aufzutauchen, dem Türausschnittprofil folgend und als Schweller endend an der Lüftungsöffnung des vorderen Radhauses. Weil zukunftsorientiert, durchaus schon mit Serienreife gesegnet!

Weniger segensreich, weil in der Designsprache von Renault bereits Standardrezept, ist die Schwellergestaltung, die sich wie eine aufgeschnittene Blechhaut gibt und Gedanken an eine Wunde mit locker gezeichneter Chromplanke verscheucht, wie besonders beim Renault Captur von 2017 zu studieren. Über dem Schweller schwelgt die Flanke in weicher, flächiger Ausbeulung – ein Motiv, das, wie beschrieben, in diesem Fall der Karosseriewerkstatt Talent abverlangt. Das sollte auch die Werkstatt haben, zu der der Ferrari 812 Superfast von 2017 zum Glattbügeln rollt. Wie eine leicht schwingende Mittelgebirgslandschaft in kleinem Maßstab steigt die vordere

motif, a fact that they realized six years later, when the line was straightened by means of a minimal trick in the rear door panel. It worked. But it didn't work in a four-door coupé and convertible version, which kept the embarrassing little detail. Quite incongruously, in the B-Class and the new 2015 A-Class, a sweeping line was driven into the metal of the flanks. It didn't match anything, neither as a curve echoing the roofline nor other motifs. But it was glossed over by the designers. For new models, design chief Gordon Wagener later coined the term »sensual clarity« – oh well, maybe it was appropriate for the new cars, but never mind about the unlucky ones that went before. And journalists like to speak of the DNA of a marque, meaning: Never deny the origin.

Incredibly, even in Russia, where there's a will there's a way – that is, if they hire a Western-trained designer – in this case Steve Mattin, a former Mercedes and Volvo designer. Lada, at the lower end of all price ranges, had come up with a concept car that no one in 2015 would have expected of this marque. It's true that the Vesta Cross Concept has a new and unusual front design, like that of the new Mitsubishis with a corrugated grille lattice. But the swages are jaunty, stamped into the metal above the front and rear wheel like a boomerang with one shorter end. Compared to previous Ladas, this is a virtual fountain of youth for the almost unmodified serial model. At the other end of the price range it is always a good idea to practice noble restraint. For instance, Rolls-Royce shifts to a lower position a line that only minimally curves upward behind the front wheel. And its former structurally identical brother, which today is manufactured by VW – the Bentley GT Speed – does something that the E-Class wasn't so good at: It copies an old 1950s Bentley fender, a practice we've often observed and described here. We are now in 2016, appropriate for a car that is not part of the normal world. Also not part of it, but only a study in 2017, is the Quant 48Volt built by the Swiss firm NanoFlowcell. This looks like a fastback luxury sedan and is supposed to reach a top speed of 300 km/h, and with an all-electric drive system at that. This is emphasized by a copper strip. Note: This is a reference to the copper wire coil of the electric motor. This strip, which is fluted above the rear window und thus follows the shape of the roof, shifts behind the panel in the flank, only to reappear below later, following the profile of the door and ending as a sill on the ventilation hole of the front wheel well. Because the design is future-oriented, it has already been slated for production!

Less serendipitous, because it is already a standard recipe in the styling of Renault vehicles, is the sill design that looks like a sheet metal skin that's been sliced open and makes one think of a wound with a carelessly drawn chrome plank in it, observed particularly in the 2017 Renault Captur. Above the sill the flank gently bulges – a motif that in this case requires talented car body makers. Such talent would also be required in the body workshop to which the 2017 Ferrari 812 Superfast is brought to be smoothed. Like a gently rolling small-scale hilly landscape the front sill rocker panels rise to the rear wheel. And because the large opening of the wheel well ventilation

231. The 2007 Opel study Flextreme deserves praise. (Photo: Opel.)

232. Before it went into series production in 2015, the Mercedes AMG A probably could have used the help of designers. (Photo: Daimler.)

233. The 2007 Hyundai HED4 is wild about glass. (Photo: Hyundai.)

234. The 2015 Lada Vesta Cross Concept takes heart after decades of dreariness. (Photo: Lada.)

235. In 2017 the Renault Captur focuses on the rocker panel. (Photo: pinthiscar.)

236. The side of the 2017 Ferrari 812 Superfast – like a mighty wave curling on the beach. Borderline acceptable. (Photo: Ferrari.)

237. The 2016 Bentley GT Speed: subtle fender lines, like in the good old days. Just right. (Photo: Bentley.)

238. The copper strip of sympathy is threaded through the bodywork of the 2017 Quant 48Volt. (Photo: Quant.)

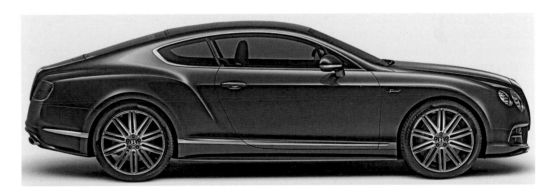

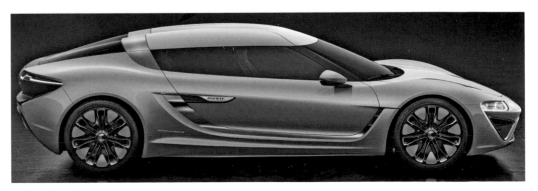

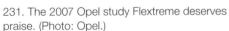

Schwellerpartie weich in die Höhe zum Hinterrad. Und weil hinter dem Vorderrad die große Öffnung der Radhausentlüftung liegt, wird diese deshalb sichtbar, da die Seitenwand der Flanke heftig eingezogen ist. So aufregend Ferraris sind, es gab schon etwas elegantere Lösungen für Leistungsportler.

Bevor das Flankenthema abzuhaken ist, muß von Rädern geredet werden. Sie sind Bestandteil der Seitenansicht in Flankenhöhe. Wo früher Speichenräder, glatte oder gelochte Stahlfelgen die Serien beherrschten, kann heute das Rad aus gegossenem Aluminium sehr frei in der Gestaltung sein nach dem Grundgesetz der gleichmäßigen Massenverteilung, damit die Zahl der Auswuchtgewichte nicht ins Bodenlose wächst. Vor einigen Jahrzehnten war der Enthusiasmus für neues Felgendesign so groß, daß bisweilen skurrile Graphiken entstanden bis hin zu unsymmetrischer Auslegung. Erst wenn das Auto mit abwegiger Felge fährt, sieht man, wie manches Rad purzelt, wenn es sich dreht. So macht keiner etwas falsch, wenn er ein Felgendesign wie das des Bugatti Royal wählt: zeitloses Aluminium.

7.3. Keine Technik, viel Form bei schnellen Linien

Ein Unterthema zu den Flanken, aber interessant genug, um gesondert abgehandelt zu werden, sind die schnellen Linien. Dieser Begriff kam irgendwann nach dem Zweiten Weltkrieg auf, als ihr Siegeszug auf den lackblanken Seiten der Karosserie begann. Keine Psychologie der Gestalt findet tiefgründige Argumente für diesen Gimmick automobilen Make-ups. Funktional gibt es keine Begründung dafür, es sei denn, geprägte Linien gäben eventuell größeren Blechflächen Stabilität, was bei damaligen Blechstärken von einem Millimeter kaum nötig war. Man denke nur an den ersten Borgward Hansa 1500 von 1949: Da flatterte nichts in den großen Pontonflanken. Und wenn der Studebaker Starliner von 1953 das Beispiel war, schnelle Linien nicht als reversiblen Schmuck, sondern dauerhaft dem Blech verbunden zu zeigen, dann sind die Applikationen mit Chromleisten oder farbig abgesetzten Flächen eher der Ausdruck des kurzfristigen Engagements in modischen Effekten. Ist dies die Erklärung, dann wundert es nicht, daß schnelle Linien nicht gerade der optische Ausweis für Hubraum sind – und für PS-starke Autos. Schnell wie Make-up sind sie aufgetragen und wieder abgeschminkt. Um im Bild zu bleiben: Die blechgeprägten schnellen Linien waren nur kleine Schönheitsoperationen.

Woher das Motiv stammen könnte, zeigt die verblüffende Ähnlichkeit der Linienführung mit Vorkriegsmodellen. Hier könnten ausschwingende Kotflügel Pate gestanden haben. Und beim Modell des Delahaye 175 M Roadster mit einer Karosserie von Saoutchik ist die Symbiose gelungen, Kotflügel und schnelle Linien in üppiger Vollendung zusammenzubringen. Handwerkliche Meisterleistung ist die flächenbündige Vermählung von verchromter und lackierter Partie auf dem vorderen und hinteren Kotflügel. Die Kotflügel deckten die Räder komplett ab, was die Wirkung der schnellen Linien, hier als vergrößerte, geschweifte Flächen, steigerte. Eine Opulenz wie diese war 1949 nicht zu erwarten. Amerikaner, inzwischen Vorbilder und Meister automobilen Designs, das damals noch Styling genannt wurde, bevor dieser Begriff den Sinngehalt änderte und damit das übertriebene, gestaltungssüchtige Entwerfen meinte. Das zeigte sich beispielsweise bei uns an der Hochschule für Gestaltung in Ulm, die klares, funktionales Design zu Weltruhm brachte und Automobildesigner als nicht standesgemäß ansah. Schade, daß es Ulm nicht mehr gibt. Gut, daß automobiles Design an Ernsthaftigkeit gewonnen hat.

239. Der Mercedes 220SE wagt es 1956 zaghaft, zweifarbig zu werden mit dem Hauch schneller Linien. Schwäbisch sparsam eben. (Photo: Daimler.)
240. 1931 nannte sich der Bugatti Royale Type 41 Coupé Napoléon. Felgen schön für alle Zeiten. (Photo: Bugatti.)
241. Die Räder sind verschieden, doch asymmetrisch gestaltet oder grob gelocht vermiesen sie die Optik beim Drehen. Könnte aber markenprägend wirken. (Photo: Autor.)
242. 1953 kann sich der Buick Roadmaster seine klassische Chromlinie auf lange Zeit in die Flanke legen. (Photo: Topcarrating.)
243. Der FSO Syrena aus Polen klaut 1957 beim 300SL und beim Studebaker Motive für die Flanke. (Photo: J. Halun.)
244. Der Wohlstand ist 1949 noch nicht ausgebrochen. Das schert den Delahaye 175S Roadster mit dem Kleid von Saoutchik herzlich wenig. (Photo: R. Kimball.)

239. The 1956 Mercedes 220SE hesitantly dares to add a second color with a hint of fast-moving lines. That's Swabian thriftiness for you. (Photo: Daimler.)
240. In 1931 the Bugatti Royale Type 41 called itself Coupé Napoléon. The wheel rims will continue to be beautiful for all time. (Photo: Bugatti.)
241. The wheels are different, but when designed asymmetrically or with large holes, their visual appearance is spoiled as they rotate. This could help shape the brand, though. (Photo: author.)
242. The sides of the 1953 Buick Roadmaster sport its classical chrome line for a long time to come. (Photo: Topcarrating.)
243. The 1957 FSO Syrena from Poland steals motifs from the 300SL and from Studebaker. (Photo: J. Halun.)
244. In 1949 prosperity is still in the distant future. But the Delahaye 175S Roadster with bodywork by Saoutchik couldn't care less. (Photo: R. Kimball.)

system is behind the front wheel, it is visible, since at that point the side wall of the flank is sharply retracted. Ferraris may be exciting, but we've had more elegant solutions for competitive sports cars.

Before leaving the topic of flanks, we must mention the wheels. They are a component of the side view at flank height. Where once spoked wheels, smooth or perforated steel rims, were predominant in serial production, today cast aluminum wheels can be very freely designed according to the basic law of the uniform distribution of mass, so that the number of balance weights will not increase inordinately. A few decades ago the enthusiasm for new rim designs was so great that there were occasional bizarre blueprints even including asymmetrical designs. It is only when the car is driving with an off center rim that we can see how poorly balanced some wheels are when they rotate. No one will go wrong when choosing a rim design like that of the Bugatti Royal: timeless aluminum.

7.3. No technology, plenty of form with fast lines

A subheading to the topic of flanks, but interesting enough to be treated as a separate category, are fast lines. This term was coined some time after World War II, when they began to appear on the blank sides of the bodywork. There is no profound psychological argument to support this gimmick of automotive makeup. There is no functional rationale for it, except for the claim that stamped lines may possibly give larger metal panels stability, something that was hardly necessary in the case of metal panels at the time, which were a millimeter thick. You've only to think of the first Borgward Hansa 1500 in 1949: Nothing was vibrating in the big pontoon flanks. And though the 1953 Studebaker Starliner exemplified a design that did not use fast lines as reversible décor, but had lines that were permanently stamped in the metal, the application of chrome strips or panels in a contrasting color tends to be the expression of a short-term commitment to modish effects. If that is the explanation, it is no surprise that fast lines are not exactly the visual proof of cylinder capacity – or prove that a car is powerful. They can be applied as quickly as makeup and removed again. To take the metaphor further: The fast lines that were stamped into the metal were merely a bit of cosmetic surgery.

The striking resemblance of the lines to prewar models shows where the model might have originated. It may have been inspired by fenders that swung outward. And in the model of the Delahaye 175 M roadster with a body by Saoutchik the symbiosis is successful – fenders and fast lines have been brought together in lush perfection. The flush-mounted combination of chrome-plated and painted sections on the front and back fender is a masterpiece of craftsmanship. The fenders completely cover the wheels, which heightens the effect of the fast lines that here appear as enlarged, curved surfaces. Opulence like this was unexpected in 1949. Americans had in the meantime become models and masters of automotive design, which at the time was still called styling, before the meaning of this term changed and it came to mean an exaggerated obsession with design. We saw this, for instance, at the Ulm School of Design, which made clear, functional design famous all over the world and looked down on automotive designers. It's too bad that the School of Design no longer exists. It's a good thing that automotive design has gained in seriousness.

In the early 1950s in the U.S. seriousness could only be expected if it meant being serious about a quick sale. Thus the fast lines of the GM's Buick marque are its special hallmark. The 1952 Buick Roadmaster Riviera still sported a fat chrome curve that traced half a recumbent parabola. By the next year the Buicks – pictured here is the convertible version »Skylark« – had the classic Buick lines, which, as explained above, trace curving fenders. Buick stayed with this

245. 1957 kann der Ford 17m die schnelle Linie in spitzer Variante mit Zweifarbenlackierung betonen. Daß er im Bug üppigen Chrom trägt, paßt. Und blechtechnisch sind Lichterschirme ein teurer Designspaß. (Photo: Ford.)
246. Die Arabella von Lloyd hat 1961 den Flankenschmuck blechgeprägt und zusätzlich chrombetont. Dazu kommt ein Panorama-Heckfenster, was in die Zeit gehört, aber wie die Heckflossen offenbar ein Muß ist. (Photo: K. Spurzem.)
247. Der Kleinste macht 1958 mit: Der Glas Isar T700 hat schnelle Linien und teure Panoramascheibe. Das Ganze gekrönt mit Zweifarbenlackierung: der US-Straßenkreuzer im Kleinstformat. (Photo: A. van Beem.)
248. Das Bedford Dormobil von 1968 kann nicht an sich halten. Der Modetrend hat sein Opfer gefunden: Weißwandreifen, Chromschmuck, schnelle Linien und Zweifarbenlackierung. (Photo: Charles01.)

Diese war Anfang der 1950er Jahre in den USA nur dann zu erwarten, wenn der Ernst des zügigen Verkaufs gemeint war. So sind die schnellen Linien der Marke Buick vom GM-Konzern ihr spezielles Kennzeichen. Der Buick Roadmaster Riviera von 1952 übte sich noch in dicker Chromkurve, eine halbe, liegende Parabel nachzeichnend. Schon ein Jahr darauf hatten die Buicks, im Bild die Cabrioversion »Skylark«, die klassische Buick-Linienführung, die, wie bereits oben erklärt, geschwungene Kotflügel nachzeichnet. Dieses Motiv behielt Buick jahrelang bei; es hatte seine größte Ausdruckskraft, als die kreisrunden Radausschnitte den Schwung der Linie noch verstärkten, wenn sie sich über dem Hinterrad dem Kreis anzupassen hatte. Bei uns durfte der Opel Rekord P1 in abgemilderter Form von 1957 bis 1960 diesen American way of life erfolgreich tragen, zusätzlich amerikanisiert mit der vorderen und hinteren Panoramascheibe. Weil beim Opel die Linie Make-up war, konnten die Designer auch eine »Billigversion« des P1 1958 lancieren. Als Opel 1200 waren alle Blechteile gleich, nur aus der kräftigen, schnellen Chromlinie wurde ein schüchterner Chromstrich, der sich Richtung Heck leicht gekrümmt neigte. Das förderte tatsächlich den (sicher nicht gewollten) Eindruck einer Billigversion. Nicht billig, aber zurückhaltend wie immer hat Mercedes den Modetrend vorsichtig begleitet. Das schon vorhandene Echo von Kotflügellinien in fast gerade geführtem Strich wurde 1956 bei dem W180 220 SE nicht nur mit feinster Chromlinie nachgezeichnet, sondern war auch zweifarbig lieferbar. Die ganz Kleinen hielten mit: Der Fram King Fulda S7 nutzte 1957 nur unterschiedliche Farbe zur Erweckung schneller Linien. So der große Ford Taunus 17M, nachgeholfen mit gezackter Chromlinie und drei kleinen Chromstegen als winziges Motiv für Tempo.

In den Varianten blechgeprägt und chromverstärkt waren viele dem Trend erlegen, der polnische FSO Syrena Sport von 1957 wie der Glas Isar T 700 von 1958, der wie ein Mini-Ami aussah. Ein Austin 55 von 1958, der Japaner Toyota Publica von 1961, die Arabella de Luxe von Lloyd aus Bremen von 1961, sie waren dankbar für die einstige Buick-Vorgabe. Dagegen stemmte sich der Wartburg 312-500 in der Campinglimousinen-Version. Hier fanden 1965 die Flanken eine ange-

motif for years and was at its most expressive when the circular wheel cutouts further intensified the momentum of the line as it conformed to the circle over the rear wheel. Here in Germany, from 1957 to 1960, the Opel Rekord P1, in an attenuated form, successfully represented this American way of life; an additional American-style feature were the front and rear panoramic windows. Since in the Opel the line was merely cosmetic, the designers were also able to launch a »cheap version« of the P1 in 1958. In the Opel 1200 all the metal parts were the same, only the strong, fast chrome line had turned into a bashful, slightly curved chrome stripe that slanted rearward. This did actually promote the (definitely unintended) impression that it was the cheap version. Not cheap, but conservative as always, Mercedes cautiously followed the fashion trend. The echo of fender lines that was already present – an almost straight line – was not only traced with a very fine chrome line in the 1956 model W180 220 SE, but the car also came in a two-tone version. The little cars followed suit: The 1957 Fram King Fulda S7 used only a contrasting color in order to emphasize its fast lines. So did the big Ford Taunus 17M, helped by a serrated chrome line and three small chrome bars, a tiny motif representing speed.

Many cars succumbed to the trend of using lines stamped into the metal and reinforced with chrome, for example, the Polish FSO Syrena Sport in 1957 and the Glas Isar T 700 in 1958 that looked like a mini version of an American car. A 1958 Austin 55, the 1961 Japanese Toyota Publica, the 1961 Lloyd Arabella de Luxe from Bremen were all grateful for features formerly cribbed

245. In 1957 the Ford 17M emphasizes the fast-moving line in an angled version with two-tone paintwork. All that chrome in the front of the model is appropriate. And from a metal technology point of view, covers for the lights are an expensive design feature. (Photo: Ford.)

246. In the 1961 Lloyd Arabella the ornamentation of the sides is stamped into the metal and is also accentuated with chrome. An added feature, the panoramic rear window, is a sign of the times, but like the tailfins obviously a must. (Photo: K. Spurzem.)

247. The smallest model joins the trend in 1958: The streamlined Glas Isar T700 has an expensive panoramic window. The crowning touch is the two-color finish: a full-size U.S. street cruiser in miniature. (Photo: A. van Beem.)

248. The 1968 Bedford Dormobil goes out of control, becoming a victim of the fashion trend: white-wall tires, chrome décor, streamlining and two-color finish. (Photo: Charles01.)

nehme, ästhetische eigenständige Form, die nur die pausbäckige Partie über den Hinterrädern störte. Ein anderer Camper aus England, das Bedford Dormobile 1968, war dagegen einfallslos mit müder, schwacher Interpretation dessen, was bei Buick einst fast zum Markenzeichen aufstrebte.

7.4. Form als Keil, Technik angepaßt

Der Keil verfolgt den Menschen seit jener Zeit, als er diesen scharfkantigen Stein aufhob. Wie scharf, spürte er, als er sich damit aus Versehen in die Handfläche ritzte. So lernte er die Schärfe sinnvoller zu nutzen als für eine Verletzung. Er schälte Rinde von Zweigen, er zog erlegtem Wild die Decke ab. Und schnell lernte er, das Teil an einen kurzen Ast zu binden, fertig war die Axt. Seither ist für uns der Keil eher ein gefährliches als ein praktisches Gerät im Alltag. Die Keilform hat etwas Aggressives, seine Aufgabe scheint das Spalten. Und wenn es die Luft ist. Es war nur logisch, diese Erkenntnis im Karosseriebau zu nutzen. Erste Ansätze waren zaghaft und wenig angespitzt. Die Karosserie neigte sich so weit, daß noch eine schmale restliche senkrechte Fläche übrigblieb, so wie es der AC Sociable von 1910 oder der Unimog 401 von 1953 zeigen. Beide keilten ihre Form nicht wegen der Aerodynamik, dafür waren sie zu langsam. Erst ab ca. 60 km/h beginnt der Luftwiderstand zum größten Fahrwiderstand zu werden. Und die ersten sportlich ausgelegten Autos nach der Jahrhundertwende nutzten lieber spitz abgerundete oder einem Bootsbug ähnliche Fronten, die einen senkrechten Keil darstellen. Im moderneren Motorsport gab es ab und zu keilförmige Karossen, die zu selten waren, um einen Trend einzuläuten. Das gelingt normalerweise nur einer Großserie, wenn sie sich auf den Straßen zeigen kann.

Wenn das ein Kriterium ist, kann der NSU Ro 80 von 1967 als erstes Beispiel für einen Serienkeil mit einer Keilform in klarer Auslegung genannt werden. Mag ein Citroën ID/DS von 1955 in der Seitenansicht dem Keil nahekommen, so fehlt ihm aber die abfallende Motorhaube bereits ab der Frontfensterwurzel, und der Bug ist kräftig ausgerundet. Hingegen macht der keilförmige Bug weniger Kompromisse und will über die gesamte Front die Keilspitze ausformen. Das macht dann, je spitzer diese Ausformung das vordere Wagenende definiert, jenen aggressiven Eindruck aus, den wir unbewußt mit der Keilform verbinden. Die erste Serienversion des Ro 80 hatte dies dadurch betont, daß die senkrechte Restfläche der Front sowohl im Grill als auch in den Abdeckgläsern der Scheinwerfer geknickt war und eine Art Nachschärfung des Hauptmotivs darstellte. Bereits nach drei Jahren stand alles wieder senkrecht ohne Knick, ohne daß der Ro 80 formal gelitten

249. 1970 wird der Keil auf den Höhepunkt getrieben. Der Lancia Stratos Zero von Bertone kennt kein praktisches Detail wie gute Rundumsicht. Dafür wird die Dreieckshaube mehrfach gefeilt und feiert die Form als Triumph. (Photo: Bertone.)
250. 1970 geht der Keil auch so. Der Ferrari Modulo 512S ist fast ein Formel-1-Renner bei vergleichsweise guten Sichtverhältnissen und perfekt ausbalancierten Flächen von Blech, Glas und Radhausöffnungen. (Photo: Ferrari.)
251. Der NSU Ro 80 von 1969 zeigt, daß sein Designer Claus Luthe ein gelernter Karosserieingenieur war und dennoch einen Klassiker schaffen konnte. Kaum ein anderes Auto kann sich bis heute formal so behaupten wie der Ro 80, der 1967 in Serie ging. (Photo: Autostadt.)

from Buick. The Wartburg 312-500 in its campervan version swam against the tide. Here, in 1965, the flanks had a pleasant, aesthetically autonomous form, disrupted only by the chubby part over the rear wheels. On the other hand, another campervan from England, the 1968 Bedford Dormobile, was unimaginative – a tired, weak interpretation of what, in a Buick, had once almost become a trademark.

7.4. Form as a wedge, technology follows suit

Wedges have been with people ever since a man picked up one such sharp-edged rock. He realized just how sharp it was when he accidentally grazed his hand with it. That's how he learned to use its sharpness for more practical purposes. He stripped the bark from branches, he pulled the hide off animals he had killed. And he quickly learned to tie the rock to a short branch, when, hey presto, he had an axe. Since then the wedge has been a rather dangerous but practical tool we use in our daily life. The wedge shape has something aggressive about it, its function appears to be to split things. Even if it's only air. It was only logical to make use of this fact for car body construction. Initial designs were tentative and not very pointed. The bodywork sloped so far that only a narrow vertical area remained, as shown by the 1910 AC Sociable or the 1953 Unimog 401. Both were not given a wedge shape because of aerodynamics, they were too slow for that. It is only upwards of 60 km/h that air resistance starts to become the greatest road resistance. And the first sports cars after the turn of the 19th century preferred to use a sharply rounded front section or one that resembled the bow of a boat, looking like a vertical wedge. In motor sports of a more recent period there was wedge-shaped bodywork, too infrequent to herald a trend. Only large-scale manufacture succeeds in doing that when it appears on the roads.

If that is a criterion, the 1967 NSU Ro 80 may be considered to be the first example of a mass-produced car with a clearly interpreted wedge shape. While the side view of a 1955 Citroën ID/DS comes close to a wedge, it lacks the sloping hood starting with the root of the windshield, and the nose is decidedly rounded. Conversely, the wedge-shaped nose makes fewer compromises and tries to extend the tip of the wedge over the entire front section. The more pointedly this shape defines the front end of the car, the stronger is the impression of aggression we subconsciously associate with the wedge shape. The first series version of the Ro 80 had emphasized this by a kink in the remaining vertical surface of the front both in the grille and in the protective glass covers of the headlights, representing a kind of sharpening of the main motif. Only three years later, everything was vertical again without a kink, and without jeopardizing the Ro 80 formally. A total of 37,406 copies of the model were sufficient to publicize the wedge shape, and the car also owes its success to the flat Wankel engine – adapted technology, so to speak.

How different, then, was the opportunity of creating the wedge in its absolute form, provided it was a concept car like the 1970 Lancia Stratos Zero. Here Bertone designed the pure wedge so uncompromisingly that the front turned out to be almost razor-sharp and the driver was practically forced to steer lying down. Sure, you could get in, but only if you first raised the windshield. There was access to the engine, but only if you first opened a huge triangular hood, the formal

249. In 1970 the wedge reaches its pinnacle. Bertone's Lancia Stratos Zero ignores practical details like good all-round visibility. On the other hand, it has improved upon its triangular hood and celebrates the form as a triumph. (Photo: Bertone.)
250. In 1970 the wedge can also look like this. The Ferrari Modulo 512S is almost a Formula 1 champion with comparatively good visibility and perfectly balanced surfaces of metal, glass and wheel well openings. (Photo: Ferrari.)
251. The 1969 NSU Ro 80 shows that its designer, Claus Luthe, was a trained automotive surface engineer, yet capable of creating a classic. There's hardly another car to this day that can hold its own formally as well as the Ro 80, which went into series production in 1967. (Photo: Autostadt.)

252. 1978 wird der Fiat X1-9 den Keil nur am Bug wollen. Scheinwerfer als Schlafaugen stören nicht das Tagesprofil. Der Entwurf stammt von Bertone. Schwarz sorgt für die nötige Schminke in den Flanken, was die Wiedererkennung fördert, nicht aber eine gute Gestaltung. (Photo: Fiat.)
253. 1974 lautet das Rezept: vorne runter, hinten hoch. So kann der VW Scirocco den Cw-Wert verbessern, ohne den Keil zu strapazieren. (Photo: Volkswagen AG.)
254. Alles ist 1984 extrem: die Länge, der Keil, die Höhe. Das darf sich nur ein Aston Martin Lagonda leisten, hier in der Ausstattungsvariante Tickford. (Photo: Aston Martin.)
255. 2017 und schon Jahre vorher hat Lamborghini den Keil fast für sich allein gepachtet. Der Huracan Performante hält sich dran und nutzt das Keilmotiv, wo er kann, wie zum Beispiel bei den Luftöffnungen. (Photo: Lamborghini.)
256. 1993 keilt sich nur der Bug des Lotus Esprit, was Schlafaugen zwingend macht. Harte Details fehlen, sie passen sich eher in gerundeter Linienführung an, was als Datum der Entstehung gelesen werden kann. (Photo: Lotus.)

hatte. Insgesamt reichten 37 406 Exemplare, um die Keilform bekanntzumachen, wofür auch dem flach bauenden Wankelmotor zu danken war, quasi als angepaßte Technik.

Wie anders war da die Möglichkeit, den Keil in seiner absoluten Form zu kreieren, vorausgesetzt es war ein Concept-Car wie der Lancia Stratos Zero von 1970. Bertone zeichnete hier den reinen Keil so kompromißlos, daß die Front fast messerscharf geriet und der Fahrer beinahe im Liegen pilotieren mußte. Einsteigen bitte, aber nur unter der hochgeklappten Frontscheibe. Zugang zum Motor bitte, aber nur nach Öffnen einer riesigen Dreieckshaube, formale Umsetzung des Keiles in der Fläche. Etwas milder gab sich der Modulo von Pininfarina. Auch aus dem Jahre 1970, war er die Camouflage für einen Formel-Renner mit fast großzügiger Verglasung und in der Seitenansicht mit dem Anflug von Symmetrie wegen seiner Öffnungen der Fenster und Radhäuser. Die Serie kann nur alles als Zitat eines Keiles verwenden. Bei dem ersten Golf von 1974 war es nur die geradlinig schräg gehaltene Motorhaube, beim Scirocco, der im selben Jahr etwas früher auf den Markt kam, förderte die Fließhecklinie den Eindruck des Keiles, besonders wenn in vergleichender Gegenüberstellung Bug und Heck Höhenunterschiede zeigten. 1978 hatte der beliebte Fiat X1/9 eine sehr zierliche, angepitzte Bugpartie, die nur dann gestört wurde, wenn die Klappscheinwerfer Licht spendeten.

Mit so einem Keil wie der Fiat, aber unter Verzicht auf versenkbare Lichter, konnte sich das Extremdesign des Aston Martin Lagonda von 1984 eine Sonderstellung bei Luxuslimousinen erlauben. Jede Fläche war so gut wie nicht bombiert, alles optisch bretthart, lange Überhänge vorne und hinten, stärker kann ein Keil nicht auffallen. Der Lagonda war nicht zu ängstlich, um in den schmalen Bug noch ein Rudiment an Kühlermaske zu integrieren. Da bleiben den echten Boliden wegen der Aerodynamik kaum übertriebenere Keile möglich. Der BMW M1 hatte, mit gerade einmal 457 gebauten Exemplaren zwischen 1978 und 1981, aus der Feder von Giorgetto Giugiaro dem Keil wie beim Golf im Frontbereich eine Chance gegeben und konnte im schmalen Bug, ähnlich wie der Lagonda, die Miniversion der BMW-Niere verewigen. Ein Lotus Esprit von 1993 und seine Vorgänger und Nachfolger hatten von 1973 bis 2006 die Keilform quasi abonniert, Giugiaro war auch hier am Design beteiligt. Überhaupt verstanden alle großen Designstudios für die schnellen Automobilvarianten den Keil als bestes Ausdrucksmittel, Power und Speed in Karosserieformen umzusetzen. Da darf auch kein Lamborghini fehlen, zum Beispiel in der Extremvariante des Huracan Performante von 2017.

7.5. Die Technik öffnet sich für die Form

»Doch die im dunkeln sieht man nicht.« Dieses Zitat von Bertolt Brecht kann einem einfallen, wenn man auf den Spuren von Flügeltüren geht. Carl Benz hatte es noch einfach mit seinem Patent. Es war die Reduzierung auf das Notwendigste zum pferdelosen Fortkommen. Eine Tür war weder gewollt noch überhaupt notwendig. Es fehlte die Wand, die man hätte öffnen müssen, um ans Volant zu kommen. Waren die ersten Ansätze von Verkleidung unterwegs, so sparte man sich immer noch Türen und schnitt lieber ein Stück so weit aus, daß mit angehobenen Beinen der Weg zum Sitz möglich war. Der offene Torpedo hatte in einigen Versionen die Tür im Heck bei komplett geschlossenen Seiten, was umständlich und nicht lange sinnvoll war. Denn Wetterschutz auch von oben brauchten Dächer. Waren sie faltbar, mußte die Seite ihre Türen bekommen. Das waren genau solche, wie sie zu Hause waren, fachmännisch schlicht Drehtür genannt. Einseitig angeschlagen, bekamen die mit hinten liegenden Scharnieren später den Spitznamen »Selbstmördertür«. Öffnete man sie unterwegs, so riß der Fahrtwind sie schneller auf, als eine rettende Hand eingreifen konnte. Es verstand sich daher von selbst, sie 1961 bei uns zu ver-

252. By 1978 the Fiat X1-9 only wants the wedge in its front section. Retractable headlights don't disturb the day profile. The design is by Bertone. Black provides the necessary highlighting on the sides and promotes brand recognition, but doesn't make for good design. (Photo: Fiat.)
253. The recipe in 1974: lower the front, raise the back. Which is how the VW Scirocco is able to improve its drag coefficient without overtaxing the wedge. (Photo: Volkswagen AG.)
254. Everything is extreme in 1984: the length, the wedge, the height. That's something only an Aston Martin Lagonda can afford. Pictured here is the Tickford version. (Photo: Aston Martin.)
255. In 2017 and for years before that Lamborghini reserved the wedge almost for its sole use. The Huracan Performante sticks to this recipe and uses the wedge motif wherever it can, for instance, in the air vents. (Photo: Lamborghini.)
256. In 1993 only the front section of the Lotus Esprit is wedge-shaped, making retractable headlights mandatory. There are no hard details: rather, they adapt in rounded styling, revealing the date the car was first manufactured. (Photo: Lotus.)

two-dimensional interpretation of the wedge. Pininfarina's Modulo was somewhat more bland. It too was built in 1970, and was camouflaged as a Formula race car with almost generous glazing, and a touch of symmetry in the profile due to the openings of the windows and wheel wells. Series production is able to use everything as an allusion to the wedge. The first Golf, in 1974, only had the rectilinear slanting hood; in the Scirocco, which came on the market somewhat earlier that same year, the hatchback line promoted the impression of a wedge, especially when you compared the front and rear and saw the differences in height. In 1978 the popular Fiat X1/9 had a very graceful, pointed front section that was only disrupted when the pop-up headlights were turned on.

With a wedge like the Fiat's, but without retractable lights, it was no wonder that the extreme design of the 1984 Aston Martin Lagonda held a special position among luxury sedans. Hardly any of the panels were cambered, everything was optically hard as a rock, there were long overhangs in front and in the rear – a wedge can't be more conspicuous than that. The Lagonda was not too timid to integrate a rudimentary radiator grille in the narrow nose. Because of the aerodynamics it was hardly possible for the really fast cars to feature more exaggerated wedges. With a mere 457 copies built between 1978 and 1981, the BMW M1, designed by Giorgetto Giugiaro, had used the wedge in the front part, similar to the Golf, and was able to place a mini version of the BMW kidney in the narrow nose, like the Lagonda. A 1993 Lotus Esprit and its predecessors and successors had, as it were, subscribed to the wedge shape from 1973 till 2006. Giugiaro was involved in designing them as well. At any rate, all the big design studios realized that for the fast types of cars the wedge was the best means of expressing power and speed in bodywork shapes. We mustn't forget the Lamborghini either, for instance, in the extreme version of the 2017 Huracan Performante.

7.5. Technology opens to let in form

»You don't see the ones who're in the dark.« This quotation by Bertolt Brecht might come to mind when you're on the trail of gullwing doors. Carl Benz still had an easy time of it with his patent. His car reduced horseless transportation to its essentials. A door was neither intentional nor even necessary. There was no wall you had to open to get to the steering wheel. Once the first suggestion of fairing appeared, you still saved yourself the expense of doors – it was easier to cut out a section wide enough to be able to reach the seat by raising your legs. The open Torpedo in some versions had the door in the rear while the sides where completely closed, which was awkward and made no sense in the long run. For you needed roofs to protect you from the weather. If the roof was retractable, there had to be doors on the sides. The doors were exactly like those you had at home, for which the technical term was simply »single-leaf door«. Hinged on one side, the ones whose hinges were at their rear were later nicknamed »suicide doors«. If you opened them en route, the airstream ripped them open faster than a helping hand could rescue you. That's why it went without saying that they were made illegal here in Germany in 1961. They were allowed again much later, but only if, as a back door, they could not be opened while the car was in motion, as is the case today with the Rolls-Royce and other models such as the 2003 Mazda RX-8. Since then, hinges have been installed only in front. Or else there is a sliding door, a welcome feature, and not only in delivery trucks. The 2012 Peugeot 1008 had one like that, which hardly proved practical in the long run. Thus maneuvering the car while reversing it and looking backwards was hardly possible when the door was open, for the open door blocked your view of the guy behind you. The 1981 BMW Z1 had a different variant of the sliding door, presumably the only one we know of that came in a standard model. In this door, or rather retractable bulkhead,

257. Der 300 SL von Mercedes geht 1954 in Serie. Schon sportliche Vorläufer des Modells machen die Flügeltüren zum weltbekannten Symbol. (Photo: Daimler.)

258. Kaum einer kannte ihn 1949, den Gomolzig Taifun. Hier schwingt die halbe Frontscheibe mit nach oben. (Photo: Pinimg.)

259. 1962 ist der Ford Cougar als Studie Nutznießer der Mercedes-Türen. Ein- und Aussteigen ist eben ein Hingucker. (Photo: Ford.)

260. Zierliche Flügelchen sind es 1972 beim BMW Turbo, kein Platz für versenkbare Scheiben. (Photo: BMW.)

261. 1976 falten sich die Flügel des Chevrolet Aerovette auf Höhe der Gürtellinie. (Photo: GM.)

262. Giugiaro entwirft 1970 den geflügelten Keil Tapiro auf technischer Basis des VW Porsche. (Photo: Car-Revs-Daily.)

bieten. Sie wurden viel später wieder nur zugelassen, wenn sie als hintere Tür während der Fahrt nicht zu öffnen waren, wie heute beim Rolls-Royce und anderen Modellen wie dem Mazda RX-8 2003. Seither sind die Scharniere nur noch vorne. Oder aber es geht auch mit einer Schiebetür, die nicht nur bei Lieferwagen einen dankbaren Abnehmer fand. Der Peugeot 1008 von 2012 hatte so eine, was sich auf Dauer wenig praktisch erwies. So war Rangieren beim Rückwärtsfahren mit dem korrigierenden Blick nach hinten bei offener Tür kaum möglich, die offene Tür deckte den Blick zum Hintermann ab. Eine andere Variante, vermutlich die einzig bekannte in einem Serienauto, hatte der BMW Z1 von 1981. Die Tür, besser gesagt ein versenkbares Schott, machte die Schwelle zum Einsteigen dadurch etwa niedriger. Dann blieb nur noch das Prinzip der Flügeltür. Noch kommen wir nicht auf den Mercedes SL 300 zu sprechen. Vielmehr gab es 1949, drei Jahre vor dem ersten Renneinsatz in Mexiko, der den SL 300 berühmt machte, ein Auto im Schatten, das den sperrigen Namen Gomolzig trug.

Dieses Auto war ein Prototyp ohne Serienchance geblieben. Sein Konstrukteur und Gestalter, Herbert Gomolzig, der auch planend im Kleinflugzeugbau tätig war, machte es damals wie viele andere auch, denn der VW Käfer bot sich bevorzugt als Ideen- und Karosserieträger an, hatte er doch ein simples, aber ingeniöses Fahrwerk mit Zentralrohrrahmen. Gomolzig dachte an eine windschlüpfige Karosserie mit leichten Ansätzen an die Tropfenform, aber ausgeprägter Haube und betonten Kotflügelkörpern, die die Scheinwerfer aufnahmen. Das Heck war in der ersten, etwas grob geratenen Version dem Kamm-Heck verwandt, also mit gekürztem Stromprofil. Das gefälligere, gut lackierte und chromgeschmückte Modell mit Fließheck sah schon wesentlich serientüchtiger aus und sollte um 25 Prozent weniger Benzin schlucken, trotz gestiegener Höchstgeschwindigkeit. Aber das alles wurde in den Schatten gestellt von Flügeltüren, die in der Dachmitte an dem längs verlaufendem Trägerprofil angeschlagen waren. Das Spezielle an den Flügeltüren: Sie nahmen beim Öffnen gleich eine Hälfte der Frontscheibe mit nach oben – eine Idee, die Jahrzehnte später für einen aufregenden Concept-Car sorgte, vermutlich aber ohne Kenntnis des Gomolzig-Autos. In den Dachpartien der Türen waren herausnehmbare Schalen, wie sie etwa eine Corvette Mitte der 1970er Jahre hatte, um etwas Targa-Gefühl zu vermitteln.

Den ganz großen Auftritt der Flügeltür hatte Mercedes nach dem gewonnenen Rennen Carrera Panamericana in Mexiko 1952. Diesen Sieg konnte auch ein in die Windschutzscheibe krachender Geier nicht verhindern. Dafür gingen Bilder um die ganze Welt, wie sich die Piloten aus dem Wagen schraubten, was durch die Flügeltüren nur bedingt leichter wurde. Die aber waren kein Gag designerischer Extravaganz. Es war die rein technisch richtige Lösung, die der Gitterrohrrahmen vorschrieb, denn der war in den Flanken recht umfangreich geworden, und so mußte die Unterkante der Tür weiter nach oben verlagert werden, was wiederum verlangte, dem Kopf mehr Freiheit zu geben. Das ging nur mit der Flügeltür, die fast bis in die Dachmitte einschnitt.

the sill was made lower for ease of access. There remained the principle of the gullwing door. We haven't even mentioned the Mercedes SL 300 yet. As a matter of fact, in 1949, three years before the SL 300 was first entered in a race in Mexico that made it famous, there was an obscure car that bore the cumbersome name of Gomolzig.

This car had remained a prototype without the prospect of being mass-produced. At the time its builder and designer, Herbert Gomolzig, who also worked as a designer in small-aircraft construction, did it like many otherfs, for the VW Beetle was a favorite source of innovation for car body designers since it had a simple but ingenious chassis with a central tube frame. Gomolzig had in mind a streamlined body with a slight hint of a drop shape, but a pronounced hood and emphasized fender systems that held the headlights. The rear in the first, somewhat crude version resembled a fastback, in other words, it had a sharply cut-off rear end. The more attractive, nicely painted and chrome-plated hatchback model looked like it had considerably more series production potential and was supposed to consume 25 percent less gas, even though its maximum speed was higher. But this was all eclipsed by gullwing doors that were attached to the longitudinal support profile in the center of the roof. The thing that was special about gullwing doors was that when you opened the vehicle they also raised up half the windshield with them – an idea incorporated decades later in an exciting concept car, but presumably without the designer being aware of the Gomolzig car. In the roof sections of the doors there were removable panels like those in a mid-1970s Corvette, giving passengers the sense of being in a targa.

The gullwing doors made their big debut after the race Mercedes won at the 1952 Carrera Panamericana in Mexico. Even a vulture crashing into the windshield could not have prevented this victory. The pictures went around the world, showing the drivers laboriously emerging from the car, which the gullwing doors made only slightly easier. But the doors were not the gag of an extravagant designer. They were the purely technical correct solution dictated by the tubular space frame, for the latter now took up a great deal of room in the flanks, and thus the lower edge of the door had to be repositioned farther up, which again required that the head had to be given more space. This only worked if you had a gullwing door, where the cutout reached almost into the center of the roof. Since 1954 the »Gullwing« Mercedes had been the epitome of the gullwing sports car, for sale to everyone (provided they had enough money). It took the car only three years of manufacture, and a total of only 1,400 specimens of the W 198, to secure the title of

257. The Mercedes 300 SL goes into series production in 1954. Already the model's sporty forerunners make its gullwing doors into a world-famous symbol. (Photo: Daimler.)
258. Hardly anyone knew the Gomolzig Taifun in 1949. Here half the front windshield swings upward. (Photo: Pinimg.)
259. The 1962 Ford Cougar study car is the beneficiary of the Mercedes doors. Getting in and out will cause heads to turn. (Photo: Ford.)
260. The 1972 BMW Turbo has dainty little wings, with no room for rolldown windows. (Photo: BMW.)
261. In 1976 the wings of the Chevrolet Aerovette fold at swageline level. (Photo: GM.)
262. Giugiaro designs a winged wedge – the 1970 Tapiro – using technology based on the VW Porsche. (Photo: Car-Revs-Daily.)

Seit 1954 war der »Gullwing«-Mercedes das für jedermann käufliche (soweit genug Geld vorhanden war) Beispiel des Flügeltür-Sportwagens schlechthin. Nur drei Jahre Bauzeit, nur 1400 Exemplare des W 198 insgesamt waren ausreichend, um sich den Titel »Sportwagen des Jahrhunderts« zu sichern. Der offene Bruder, seit 1957 in der Roadsterversion, hatte dann normale Türen, weil weder Dach noch Gitterrohrrahmen diese Flügeltürtechnik notwendig machten. Offen kam der SL auf 1858 gebaute Einheiten. Beide sind heute extrem teuer und ständig auf der Suchliste bei Oldtimerauktionen. Kleines Detail beim Coupé: Das Einsteigen wurde nicht dank der Türen leichter, auch das Lenkrad konnte dafür abgeklappt werden.

Allein der Schattenriß des Autos mit offenen Flügeltüren hat Wiedererkennungswert, wie der Mercedes-Stern. Und von dieser Kraft des Besonderen scheinen sich viele einen Teil zu retten, unabhängig davon, ob sie für einen unkomplizierten Einstieg notwendig waren. Der Ford Cougar 1962 war ein Konzeptauto, das nebenbei wegen der Flügeltüren etwas vom Glanz eines SL 300 profitieren wollte. Auch der BMW Turbo von 1972 mag ähnlichem Trieb gefolgt sein, denn eine normale Tür wäre für einen akzeptablen Zugang ans Volant vorstellbar gewesen. Auffallend aber waren die sehr dünnen Türen und die sparsame innere Verkleidung. Noch dünner, weil schieres Glas, machte es der Panther Ferrari 365 GTB 4 Shooting Brake von 1975. Das ideale Einkaufsauto: einfach Klappe auf, Tüte rein, weiter shoppen. Eine Idee, die der kleiner Fiat-Konzeptwagen Vanessa schon 1957 zeigte. Viel aufwendiger als ein Glasflügel waren die Türen des Chevrolet Aerovette von 1976, hier knickte der Flügel zusätzlich nochmals mittig ein, was versenkbare Fenster nicht oder nur mit hohem Technikaufwand möglich machen könnte. Nicht anders war es beim Ford Corrida von 1976, der dafür auf versenkbare Scheiben verzichtete. Der Porsche Tapiro 1970 und der Ford Corrida von 1976 bekamen auch ihre Flügeltüren, als könne ein Sportwagen, egal mit welchem Hubraum, die Passagiere nicht anders hereinlassen. Immerhin ist zu bedenken, daß die Karosseriesteifigkeit leidet, wenn Türen in das Dach schneiden, weil so die Rahmenwirkung der Karosserieseiten unterbrochen wird. Ergo muß für eine verwindungssteife Karosserie die Bodengruppe, ähnlich wie bei Cabrios, verstärkt werden.

Wie mutig sind da Amateurlösungen! Das Beispiel des russischen GT 77 auf Skoda 1202-Basis und mit Teilen eines Moskwitsch 412r ist deshalb lobenswert, weil sein Erbauer Ordyan Sarki-

263. 1986 bleibt Giugiaro seiner Handschrift treu, auch wenn unter dem Incas ein Oldsmobil die Technik liefert. (Photo: Italdesign.)
264. Handwerkliche Einzelleistung glückt 1977 auch formal: der GT-77 Skoda 1202-Moskwitch 412R von R. Ordyan Sarkisian. (Photo: Ordyan.)
265. Der Bulldog von Aston Martin kann 1980 das Extreme des Lagonda nutzen. Flächen werden brutal glatt und ungekrümmt. (Photo: Aston Martin.)
266. Bertone entwirft 1994 auf Porsche-Basis den Karisma, ein Name, den auch Mitsubishi bei einem Langweiler nutzte. (Photo: Favcars.)
267. 2004 entscheidet sich der Bristol Fighter für ein übergroßes Heckfenster und Flügeltüren, beides gerade noch erträglich. (Photo: Phacemag.)
268. Eberhard Schultz bewies 1984 mit seinem Isdera Imperator, daß perfektes Design keinen Großkonzern braucht. (Photo: Isdera.)

263. In 1986 Giugiaro stays true to his signature style, even though under the Incas an Oldsmobile provides the technology. (Photo: Italdesign.)
264. A stand-alone achievement of craftsmanship is also a success from a formal point of view: the 1977 GT-77 Skoda 1202-Moskvich 412R of R. Ordyan Sarkisian. (Photo: Ordyan.)
265. Aston Martin's 1980 Bulldog uses the extreme features of the Lagonda. Surfaces become brutally smooth and straight. (Photo: Aston Martin.)
266. On a Porsche wheelbase, Bertone designs the 1994 Karisma, a name also used by Mitsubishi for one of its nerdy models. (Photo: Favcars.)
267. In 2004 the Bristol Fighter opts for an over-sized rear window and gullwing doors, both just barely tolerable. (Photo: Phacemag.)
268. Eberhard Schultz's 1984 Isdera Imperator proved that perfect design doesn't need a major manufacturer. (Photo: Isdera.)

»Sports Car of the Century«. Meanwhile its open brother, in a roadster version since 1957, had normal doors, because neither the roof nor the tubular space frame necessitated this gull wing technology. A total of 1,858 SLs were built. Both the W 198 and the SL 300 are extremely expensive today and are constantly on the search list at old-timer auctions. A small detail in the coupé: Getting in got easier, no thanks to the doors, and the steering wheel, too, could be folded down for easier entry.

The silhouette of the car alone, with open gullwing doors, is a brand recognition factor, like the Mercedes star. And many people seem to want to share part of the power of this special feature, regardless of whether the doors were necessary for an uncomplicated entry. The 1962 Ford Cougar was a concept car whose design incidentally included gullwing doors because it wanted to profit a little from the glamor of an SL 300. The 1972 BMW Turbo may have followed the same instinct, for it is conceivable that a normal door would have provided acceptable access to the steering wheel. Its very thin doors and the spare interior paneling were striking. Made of sheer glass, these features were even thinner in the 1975 Panther Ferrari 365 GTB 4 Shooting Brake. It was the ideal car for shopping: simply open the tailgate, toss in the paper bag, go on shopping – an idea that a small Fiat concept car, the Vanessa, had already demonstrated in 1957. Far more costly than a glass wing were the doors of the 1976 Chevrolet Aerovette: Here the wing also had an additional kink in the center, making it impossible (or only possible at great technical expense) to have windows that could be lowered. The same was true of the 1976 Ford Corrida, which therefore did not have lowerable windows. The 1970 Porsche Tapiro and the 1976 Ford Corrida also got their gullwing doors, as though a sports car, regardless of its engine capacity, could not let its passengers in any other way. Still, it should be noted that body rigidity suffers when doors cut into the roof, because the structural frame effect of the body sides is interrupted. That is why in order to have a tortionally rigid body the floor assembly, just as in convertibles, must be reinforced.

How courageous amateur solutions are in comparison! The example of the Russian GT 77 on the base of a Skoda 1202 and with the parts of a Moskvitch 412r is commendable, because its

sian 1977 mit reduzierten Möglichkeiten eine passable Karosserie schuf – auch dieses Auto lebte seither vom Imagevorbild SL 300. Daß ein Auto eines professionellen Herstellers mit Flügeltüren nicht schöner werden muß, macht der Bulldog von Aston Martin klar. Ein Auto von 1980, das in allem extrem war, konnten Flügeltüren da noch nachhelfen? Oder waren sie eine Notwendigkeit für ältere Kunden, denen man die klassische Tür nicht mehr zumuten wollte, um mit der Flügelvariante beim Einsteigen Erleichterung zu verschaffen? Übrigens waren die Scheiben auch hier nicht versenkbar, wie bei (fast) keinem, auch nicht beim DeLorean von 1981, der deshalb weltbekannt wurde, weil er als futuristisch umgeschweißtes Monster im Film *Zurück in die Zukunft* auftauchte und vom Alltag wegtauchte, weil er zu schwach motorisiert und technisch unzuverlässig war. Der Kodiak F1 konnte die Flügeltüren aus Imagegründen und zum erleichtertem Einstieg genauso nutzen wie 1984 der Isdera Imperator des Designers Eberhard Schulz. Dieser Entwurf wie auch der CW311 und der Commendatore sind trotz kleinster Auflagen im Design auf dem Niveau großer Karossiers, wie sie Italien zahlreich vorzeigen konnte. Der Oldsmobile Incas von 1986 hat das kleine Detail, das die Frage beantwortet, wie man die Tür schließen kann, wenn man bereits Platz genommen hat. Die einen bieten große Griffe oder in Griffnähe liegende Armlehnen an, der Incas, ein Giugiaro-Design, nutzt Schlaufen: billig, einfach und praktisch. So wie einst der Käfer Anfang der 1950er Jahre, hier als Griffhilfe, wenn man sich von der Rückbank hervorhangelte. Das Türschließen verlangt dann eine geometrische Untersuchung in der Mischung Greiflänge, Grifflage und Türhöhe.

VW Käfer sollen es bei Privatschweißern auch zur Flügeltür geschafft haben, aber in perfekter Concept-Car-Qualität zeigt der VW Futura von 1989 das Thema ganz in Glas, gefaßt in kräftigem Rahmen. Es muß wohl immer die Dramatik sein, die das Schauspiel des Öffnens bietet, was leicht ins Lächerlich kippen kann, wenn das Auto ein kleiner Wagen ist. Dies kann am Mazda HRX von 1991 studiert werden. Daß ein Auto den Namen Charisma trägt, wirkt bei dem kleinen Mitsubishi Carisma leicht peinlich, bei dem Porsche Karisma von Bertone aus dem Jahre 1994 ist das eher angemessen. Neben der dicken Innenverkleidung ist erkennbar, daß im großen Fenster kleine Ausschnitte sind. So gibt es die Möglichkeit, entweder Schiebefenster einzusetzen oder die Fenster zu versenken, um wenigstens etwas Luft hereinzulassen oder nach dem Weg fragen zu können, ohne mit den Flügeln zu schlagen. Der Opel X-Treme trug 2001 seine Flügel wahrscheinlich gerne, weil er als Extremsportler einen Innenkäfig hat, der das Einsteigen ohne sie wesentlich schwerer machte. Flügel beim Infiniti Triant, einem SUV-ähnlichen Gefährt, sind eben der Gag mit Drang zur Aufmerksamkeit, seine Bauhöhe macht das nicht notwendig. Beim Bristol Fighter von

builder Ordyan Sarkisian, in 1977, with limited capabilities, created a passable body – this car too ever after rode on the coattails of the SL 300, on whose image it was modeled. A car built by a professional manufacturer doesn't have to be more beautiful because it has gullwing doors, as the Aston Martin Bulldog demonstrates. Built in 1980, the car was extreme in every respect – could gullwing doors improve it? Or were they a necessity for older customers who could no longer be expected to use the classic door, whereas the gullwing version made it easier to get in the car? Incidentally, here too the windows could not be lowered, as in (almost) none of them, not even in the 1981 DeLorean, which became world famous because it appeared in the film *Back to the Future*, and disappeared from everyday use because it was too weakly motorized and technically unreliable. The Kodiak F1 was able to make good use of the gullwing doors for image reasons and for ease of access just like the 1984 Isdera Imperator by designer Eberhard Schulz. This design, as well as that of the CW311 and of the Commendatore are, in spite of very small production runs, on the same niveau as the designs of the great bodywork builders, of whom Italy has so many. The 1986 Oldsmobile Incas has the small detail that answers the question of how to close the door when you're already seated. Some manufacturers provide big handles or armrests within reach, while the Incas, a Giugiaro design, uses straps: cheap, simple and practical. Just like the Beetle's in the early 1950s, here it helps you to hold on as you get out of the backseat. Closing the door then depends on a study of several factors: how high you can reach, the position of the handle and the height of the door.

It is said that there have been VW Beetles that have had gullwing doors installed by amateur welders; meanwhile, meeting perfect concept-car quality standards, in the 1989 Futura this is an all-glass feature, housed in a robust frame. There's always a great deal of drama involved in opening the doors, which can easily verge on the ludicrous if the car is small. We can see this in the 1991 Mazda HRX. The name charisma is rather embarrassing for a car like the little Mitsubishi Carisma, while it is more appropriate for Bertone's 1994 Porsche Karisma. In addition to thick interior paneling it is apparent that there are small cutouts in the large windows. Thus you have the possibility either of installing sliding windows or of lowering the windows in order to at least let in some air or ask for directions without beating your wings. The 2001 Opel X-Treme, an extreme sports car, probably liked its wings, because it has an interior cage that made entering the car without them much more difficult. Wigs in the Infiniti Triant, an SUV-like vehicle, are simply a gag to attract attention: its height does not require them. In the 2004 Bristol Fighter a flat body resorts to wings in order to spare the British gentleman driver athletic exertion. In the GDR there was a sports car on a Wartburg platform. After the Wall came down, the Trabbi and the Wartburg went

269. In the 1989 study Futura, VW uses gullwing doors for the van version. (Photo: Volkswagen AG.)
270. Even without gullwing doors, the 2007 Mazda study Ryuga is convincing, because the designers have come up with new ideas. (Photo: Mazda.)
271. The gullwinged 2013 Buick Riviera concept has a waterfall grille. (Photo: GM.)
272. The 2006 Melkus 2000 RS comes from the newly formed German states; its precursor was a hit on East German racetracks. (Photo: Pinimg.)
273. Tesla's 2015 Model-X is like a catalog of doors: a tailgate in the rear, a revolving door in front, gullwing door in the back. It costs a bundle. (Photo: Tesla.)
274. The 2003 Infiniti study Triant makes up for its plain face with its gullwing doors. Always helpful. (Photo: Infiniti.)

2004 greift seine flache Karosse zu Flügeln, um dem englischen Herrenfahrer sportlichen Einsatz zu ersparen. In der DDR gab es einen Sportwagen auf Basis des Wartburg. Nach dem Fall der Mauer kamen Trabbi und Wartburg unter die Räder, der Melkus aber lebte auf als RS 2000. Das Design stammt von Lutz Fügener, der aus den neuen Bundesländern nach Pforzheim zum Professor für Transportation Design berufen wurde. Das Vorläufermodell Melkus RS1000 hatte bereits 1969 die Flügel, was auf den Straßen von Leipzig oder Dresden Furore machte. Fast inflationär sind jetzt bei Konzeptmodellen Flügel verwendet worden – der Mazda Ryuga von 2007 fiel mehr durch seine fein gegliederte Flanke auf als mit seinen Flügeln. Der Nuvis von Hyundai hatte 2009 als Konzept riesige Türen. Der Mercedes SLS ist 2009 eine serienmäßige und gelungene Reverenz an den SL 300. Der Buick Riviera trägt einen klassischen Buick-Namen und war 2013 eine für amerikanische Verhältnisse sehr harmonische Studie. Sein Landsmann machte 2015 im Tesla-Modell X die Flügel SUV-tauglich und begründete sie mit vereinfachtem Einsteigen auf beengten Parkplätzen (wie man sie bei vielen Supermärkten der Welt findet). Der Serienstart bescherte Tesla einige Qualitätsprobleme, die auch die Flügeltüren hatten. Und noch nicht Serie, aber kraftvoll elektrisiert gibt sich der Quant 48Volt aus der gemächlichen Schweiz, gut für 300 km/h. Aber die Größe solcher Flügeltüren verlangt, ähnlich dem Nuvis, eine starke Struktur im mittleren Dachholm. Gerade Serienmodelle haben hier erheblich umfangreichere Innenverkleidungen, die das Türgewicht ansteigen lassen. So muß auch die Konstruktion der Teleskopstangen entsprechend ausgelegt sein, um das schwebende Gewicht vor dem Zurückfallen zu sichern. Sollte noch eine elektrische Schließmechanik das Greifen nach der hochliegenden Tür vereinfachen, kommt manches Kilo dazu. Aber alle mit wenn und aber versehen und alle hier gezeigten oder weiterhin aufgezählten beflügelten Modelle haben nur einen Meister: den SL 300.

7.6. Technik und Form erlauben negative Heckfenster

Zur Geschichtsschreibung sucht man Anfänge, erste Beispiele. So auch bei der Suche nach Autokarosserien mit negativ stehenden Heckfenstern. Bevor man fündig wird, fragt man sich: Wann ist es sinnvoll, die Heckfenster schräg nach innen zu stellen? Triviale Antwort: im Winter: Der Schnee bleibt nicht liegen, die Sicht nach rückwärts bleibt frei. Weitere triviale Antwort: Der Regen fällt nicht auf das Glas, kein großer Vorteil. Triviale Antwort bei Sonnenschein: Die Aufheizung ist wegen der verschatteten Fenster geringer. Noch ein möglicher Gedanke: Der Aufbau ist auf knappe Länge reduziert, denn die Heckscheibe hat dieselbe Neigung wie die Rückenlehne der Fondsitze. Dies alles sind kaum Vorteile – aerodynamisch gibt es wahrscheinlich noch den Nachteil von Verwirbelungen –, denn sonst hätte sich diese Formgebung eher durchgesetzt. Selbst der Kostenvorteil durch plane Glasflächen überzeugt nicht, weil Autos auch in Zeiten kutschengeprägter Gestaltung ungewölbte Scheiben hatten. So bleibt für das Auge nur der Effekt des Ungewöhnlichen übrig. Dieses Motiv, ungewöhnlich zu sein, prägt sicher viele Beispiele automobiler Formgebung.

So brachte die Suche nach einem Auto mit negativ geschrägter Heckscheibe 1930 den Rolls-Royce Phantom I Windblown Brewster ans Licht. Sein Name »Windblown« ist korrekt. Der ganze Oberbau neigt sich mit ständig parallel geführten Fenstersäulen nach hinten bis zur negativ stehenden Heckscheibe, als hätte der Wind ein labiles Dach in die Schräge geblasen. Die Karosserie stammt vom US-Autobauer Brewster, der neben Rolls-Royce-Kleidern auch eigene Fahrzeuge herstellte. Ihre Eigenart bei sonst konservativer Form: Der Kühler lief nach unten zur Mitte spitz aus, begleitet von den Innenseiten der Kotflügel. Ein weiterer Rolls-Royce nahm sich in der Version des Silver Wraith die Freiheit des besonderen Heckfensters. 1954 nutzte die italienische Karosserieschmiede Vignale dessen Fahrgestell und hatte neben dem schrägen Heckfenster einen Bug, der sich nicht entscheiden konnte, italienisch elegant oder englisch konservativ zu sein. in

275. Ford betont 1958 mit seinem Lincoln Continental MkIII Landau den Anspruch, am häufigsten von allen Marken negative Heckfenster zu bauen. (Photo: Ford.)
276. 1930 ist der Aufbau vom Winde verweht. Deshalb heißt der Rolls Royce Phantom I Brewster vielleicht auch »Windblown«. (Photo: P. Durkin.)
277. Der Ford Mercury D-528 hat 1955 ein negatives Heckfenster. Aber sein Namenszusatz »Beldone« paßt so gar nicht zu den Heckflossenbeulen, nur um die Reserveräder zu verstauen. (Photo: Petersen.)
278. 1954 verging sich Vignale am Rolls Royce Silver Wraith. Was am Heck beim Fenster gerade noch verkraftbar ist, am Bug war es weder englisch noch italienisch elegant. (Photo: Dukketeater.)

275. With its 1958 Lincoln Continental Mark III Landau, Ford emphasizes its claim that out of all brands it builds the largest number of reverse-slanting rear windshields. (Photo: Ford.)

276. In 1930 the body is exposed to the wind. Maybe that's why the Rolls Royce Phantom I Brewster is called »Windblown«. (Photo: P. Durkin.)

277. The 1955 Ford Mercury D-528 has a reverse-slanting rear window. But its additional name, »Beldone«, really doesn't go with the tailfin bumps, which are there merely to house the spare tires. (Photo: Petersen.)

278. In 1954 Vignale went to work with a vegeance on the Rolls Royce Silver Wraith. While what he did in the rear by the window is only barely acceptable, the front section showed neither British nor Italian elegance. (Photo: Dukketeater.)

to the dogs, while the Melkus got a new lease on life as the RS 2000. The design is the work of Lutz Fügener, an East German who was appointed professor for transportation design in Pforzheim. Even in 1969 its precursor, the Melkus RS1000, already had the gullwing doors that caused such a stir on the streets of Leipzig or Dresden. There is now almost an inflationary use of gull-wing doors – the 2007 Mazda Ryuga stood out more due to its finely structured flank than its wings. The innovative concept of Hyundai's 2009 Nuvis was its huge doors. The 2009 Mercedes SLS is a production car, a successful bow to the SL 300. The 2013 Buick Riviera bears a classic Buick name and was a very harmonious study even by American standards. Its compatriot, Tesla's 2015 Model X, adapted the wings for use in an SUV, the rationale being that they made it simpler to get in and out in cramped parking spaces (like those found next to supermarkets worldwide). The fact that it started out as a production car meant that Tesla had a few quality issues, issues that gullwing doors also had. Not yet a production car, but powerfully electrified, is the Quant 48Volt from laid-back Switzerland, good for 300 km/h. But the size of such gullback doors requires, as in the Nuvis, a strong structure in the central roof pillar. Here, series models especially have considerably more extensive interior paneling, which increases the weight of the doors. Thus the construction of the telescopic rods must be designed so as to ensure that the floating weight does not drop back down. If there is also an electric closing mechanism to simplify one's reaching up for the door, this adds quite a few pounds to the weight. One car, though, ranks above all the rest of the winged models shown here or listed subsequently: the SL 300.

7.6. Technology and form make negative rear windows possible

When we write history, we look for beginnings, first examples. This is also the case when we look for bodywork with negative rear windows. Before we find them, we ask ourselves: When does it make sense to position the rear windows so that they are inclined inward? The trivial answer: in winter. Snow doesn't stick, the rear view always remains unobstructed. Another trivial answer: The rain doesn't fall on the glass, not a big advantage. A trivial answer on sunny days: It takes less to heat the car because the windows are shaded. One more possible idea: The bodywork is reduced to a compact size, for the rear window has the same slope as the backrest of the backseats. All of these are scarcely advantages –, from an aerodynamic point of view there is probably also the disadvantage of turbulence – for otherwise this design would have been accepted sooner. Even the fact that flat glazed surfaces are less expensive is not convincing, because even in the era when design was influenced by carriages, cars did have flat windows. There remains the argument that our eyes are struck by unusual effects. That fact is surely the reason for many examples of automotive design.

den USA finden wir den Ford Mercury D528 Beldone von 1955. Seine chromlose Bescheidenheit der Flanken wird fragwürdig, denn keiner bewundert das schräge Heckfenster, stört sich aber an den Beulen der Heckkotflügel, die offensichtlich den Reserverädern Platz schaffen sollem. Der hubraumkleine Fiat 1100 Modell 103 Coupé vom Karossier Sartorelli hätte als Zweisitzer Platz genug für eine große Heckscheibe à la Studebaker Starliner, frönt aber 1957 dem neuen Modetrend der negativ gestellten Heckscheibe, verzichtet auf Chromschmuck und hält die Heckflossen im Zaum, was den Blick auf ebendiesen Trend lenkt.

Ab 1958 wird Ford fast zum Universalerben der negativen Heckscheiben. Das Konzeptauto Ford La Galaxie nutzt dieses Motiv im vorderen Flankenbereich, wo eine geriffelte Chromfläche das Profil der hinteren Dachpartie ankündigt, alles gestreckt in endlos langer Seitenansicht. Diese bereicherte noch im selben Jahr die Serie. Hier hatte der Ford Lincoln Continental Mark III Landau einen Modellnamen so lang wie das Auto und war Ausdruck des steigenden Wohlstands. Selbst die Farbe Rosa hatte die Chance zu einer Lackierung, ohne anstößig zu wirken. Im damaligen Italien gab es viele kleine und große Karosseriebetriebe und Designer, die neben vorbildlichen Entwürfen auch amerikanische Rezepte ausprobierten und mit Handwerkskunst schnell ein Motiv in Blech umsetzen konnten. Das machte Pininfarina 1959 auf den Rädern des Fiat 600 Coupés. Ein starkes Heckfenstermotiv verstört aber die ruhige Seitenansicht mit einem hinteren Seitenfenster in der ungewohnten Form des umgedrehten Trapezes. So blieb dieses Wägelchen kein erwähnenswerter Meilenstein italienischer Gestaltungskunst. Dafür machte Ford die neue Form des Dachabschlusses auch in Europa bekannt. Die englischen Fordwerke schickten 1960 den Ford 10 SE Anglia mit allen Varianten vom Band, wie 1961 den Ford Consul Classic. Autos, auch in kleiner Zahl auf deutschen Straßen, zeigten, daß man negative Heckfenster verkaufen und im Falle Ford zum Markenzeichen werden lassen konnte.

Frankreich, immer gut für automobile Überraschungen technischer und gestalterischer Lösungen mit dem Höhepunkt des genialen Citroën ID/DS (ausgesprochen und übersetzt: »Idee« und »Göttin«, als beste Beschreibung in einem Wort) hatte bei Citroën 1961 mit einem Auto mit der Federungstechnik und dem Motor der »Ente« einen würdigen Bruder präsentiert. Mit dem speziellen Heckfenster und einem zerknautschten Gesicht hielt sich der Erfolg in Grenzen und trieb manchen zur Frage, warum das Ende eines Daches so sein muß, auch wenn erwähnte Gründe dafür sprechen mögen. Ein Mazda Carol von 1962 wollte gar in winziger Karosse den Duft der weiten Ford-Welt atmen und sah dennoch nur japanische Straßen. Bald schien das Ende dieser Fenstermode gekommen. Viel später griff der Toyota Will Vi die Idee in seiner Konzeptstudie 2000 noch einmal auf. Jetzt warten wir auf die Wiederentdeckung dieser Fensteridee.

7.7. Die Technik verschieden, die Form ein Vorbild aus den USA

Auf dem Höhepunkt von Flossen und Panoramascheiben schickte GM 1960 ein für seine Verhältnisse kompaktes Auto auf die Straßen. Die Hausmarke Chevrolet baute den Corvair, ein Design

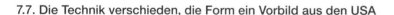

279. 1960 ein weiterer Ford: der Anglia 10 5E. (Photo: A. van Beem.)
280. Das Fiat 1100 103 Coupé von Sartorelli macht 1957 unmotiviert die Mode mit. (Photo: Sartorelli.)
281. 1958 noch ein Ford: die Studie La Galaxie. Die Länge wird durch das Heckfenster zusätzlich betont. (Photo: Ford.)
282. 1959 hilft das Heckfenster des Fiat 600 Coupé von Pininfarina, die Kürze des Autos zu mildern. (Photo: Wikimedia.)
283. Der Winzling Mazda Carol putzt sich 1962 heraus: mit viel Chrom und eben dem Heckfenster. (Photo: Mazda.)
284. 2000 sieht die Toyota-Studie Will VI wie ein Spielzeugauto aus. Das liegt nicht am Heckfenster, aber an Blechdetails. (Photo: Toyota.)
285. Citroën nutzt 1961 seine Extravaganzen, hier am Ami 6. Praktisch fraglich, doch immer auffallend. (Photo: An-D.)

279. One more Ford: the 1960 Anglia 10 5E. (Photo: A. van Beem.)

280. The 1957 Fiat 1100 103 coupé by Sartorelli follows the trend for no apparent reason. (Photo: Sartorelli.)

281. Yet another Ford: the 1958 study La Galaxie. The length is accentuated by the rear windshield. (Photo: Ford.)

282. The rear windshield of Pininfarina's 1959 Fiat 600 coupé helps to mitigate the fact that the car is so short. (Photo: Wikimedia.)

283. The diminutive 1962 Mazda Carol gets spruced up: with lots of chrome and the very same rear windshield. (Photo: Mazda.)

284. The 2000 Toyota study Will VI looks like a toy car. That isn't due to the rear windshield but to the bodywork details. (Photo: Toyota.)

285. The 1961 Citroën capitalizes on its extravagances. Pictured here: the Ami 6. In use dubious, but definitely eye-catching. (Photo: An-D.)

In one such example, the search for a car with a negatively inclined rear window produced the 1930 Rolls-Royce Phantom I Windblown Brewster. The name »Windblown« is accurate. The entire top section, with window pillars kept constantly parallel, slopes backward to the negative rear window, like a dilapidated roof blown by the wind till it leans to one side. The bodywork is by the American automaker Brewster, who also built his own cars in addition to Rolls-Royce bodywork. Otherwise conservative, their special feature lay in the fact that the lower part of the radiator tapered to a central point, following the outlines of the inner sides of the fenders. Another Rolls-Royce, the Silver Wraith, also had the special rear window. In 1954 the Italian car body factory Vignale used its chassis and in addition to the inclined rear window had a front section that couldn't make up its mind whether to be elegant like its Italian counterparts or conservative like its British brothers. In the U.S. we have the 1955 Ford Mercury D528 Beldone. The chromeless modesty of its flanks tends to be suspect, for no one admires the slanting rear window, while people are bothered by the bulging rear fenders, obviously intended to create room for the spare tires. The small-capacity 1957 Fiat 1100/103 coupé by the auto body builder Sartorelli, a two-seater, would have enough room for a large rear window à la Studebaker Starliner, but indulges in the new fashion trend of the negatively sloping rear window, is not chrome-plated and keeps its tailfins to a minimum, which focuses attention precisely on this trend.

After 1958 Ford becomes almost the sole heir of the negative rear windows. The concept car Ford La Galaxie uses this motif in the front section of the flanks, where a corrugated chrome surface announces the profile of the rear roof section, all of it elongated in an endlessly long side elevation. The latter enhanced the series that same year. Here the Ford Lincoln Continental Mark III Landau had a model name as long as the car itself and was an expression of growing prosperity. Even the color pink got a chance to be chosen for the paintwork without looking objectionable. At the time there were many small and large car body factories and designers in Italy who, beside exemplary designs, were also trying out American recipes and were adept at quickly implementing a design in metal. In 1959 Pininfarina worked on the wheels of the Fiat 600 coupé. However, a strong rear window motif throws out of balance the peaceful profile with a rear side window in the unaccustomed form of an inverted trapezoid. Thus this little car was not to be a noteworthy milestone of the art of Italian design. On the other hand, Ford also popularized the new form of the rear edge of the roof in Europe. In 1960 the Ford 10 SE Anglia left the assembly lines of the British Ford plants with all its variants, and in 1961 the Ford Consul Classic was launched. Cars, even though their number was small on German roads, demonstrated that you could sell negative rear windows and that in the case of Ford these could become a trademark.

In the 1961 Citroën, France, always good for automotive surprises involving technical and design-related solutions – the culmination being the brilliant Citroën ID/DS (pronounced and translated as »idea« and »goddess«, its best one-word description) – had a car with the suspension technology and the engine of the »Duck« that was the Ford's worthy brother. With its special rear window and a crumpled face, its success was limited and caused many people to ask why the end of a roof had to be that way, even if there was a good reason for it. A 1962 Mazda Carol, which had a tiny body, even wanted to get a whiff of the big, wide world of Ford, yet got to see only Japanese roads. Soon the end seemed to have arrived for this window fashion. Much later the 2000 Toyota Will Vi again took up the idea in its concept study. Now we are waiting for this window idea to be rediscovered.

von strenger Disziplin. Klassische Frontscheibe, gewölbt, aber nicht Panorama, eine Heckscheibe wie bei großen GM-Modellen, positiv stehende C-Säule und Dachüberstand, wie sie unser Opel Kapitän ein Jahr zuvor wenig elegant zeigte. Das Merkmal des neuen Designs war: die komplett umlaufende Gürtellinie chromgefaßt, an der Front zwischen den Lichtern leicht abgesenkt, um den fehlenden Grill zu ersetzen. Der Eindruck einer Badewanne für den Unterbau ist naheliegend. Der fehlende Grill wurde wettgemacht durch das Logo auf Chromstreifen und verwies so auf den im Heck verbauten Motor, der nur wenige Haubenschlitze zum Atmen brauchte. Keiner ahnte, daß dieser Entwurf fahrtechnisch ein Versager war wegen des unzureichenden Fahrwerks und der Hecklastigkeit, aber trotzdem zu einer der meistkopierten Vorlagen wurde, oft in Europa, seltener in Japan. Allen gemein war die umlaufende Linie mit ihrem funktionalen Vorteil: Der gesamte Unterbau wurde in ein diszipliniertes Volumen gezwungen. Und der Aufbau bot dank schlanker Dachsäulen einen Rundumblick, wie ihn heute kein Auto mehr bietet – mit der Entschuldigung, daß passive Sicherheit heute wesentlich stärkere Säulen braucht und teilweise Airbags aufzunehmen hat, meist aber auch Designereinfälle verkraften muß.

Die mußte der Fiat 1300/1500 als erster Corvairlinien-Adept bereits 1961 nicht fürchten. Fast hundertprozentig übernahm dieser die umlaufende Linie, die spiegelbildlich zur Front auch im Heck geringfügig abgesenkt wurde. Doppelscheinwerfer und identisch geformte Heckfenster ergänzten die Kopierleistung, die beim Fiat wegen der klaren Auslegung in der Seitenansicht auch ein Vertreter der Trapezlinie war – einer Stilrichtung, die besonders Pininfarina bei seinen Auftragsarbeiten für Lancia und Peugeot in diesen Jahren anzuwenden pflegte. Ein unauffälliges Design-Detail des Fiat waren die schlanken Türgriffe, die den umlaufenden Chromstreifen nutzten, um sich in ihnen dank ihrer Schlankheit zu verstecken. Nicht direkt kopiert, aber inspiriert vom Corvair, hatte BMW 1962 mit dem italienischen Designer Giovanni Michelotti den 1500 gestaltet. Der Aufbau war klassisch ohne das Panoramaheckfenster mit Dachüberstand. Der umlaufende Chromstreifen beschränkte sich auf die Heckpartie. Dafür wurde die umlaufende horizontale Linie als Sicke und als Trennfuge für Front- und Heckhaube genutzt, die Türgriffe wurden in die Sicke gelegt. Alles zusammen schuf dies einen sehr geschlossenen Eindruck. Und hier wurde das erste Mal der sogenannte Hofmeister-Knick gezeigt: der kleine negative Einzug am hinteren Seitenfenster knapp über der Gürtellinie. Wilhelm Hofmeister war seinerzeit der BMW-Chefdesigner (von 1955 bis 1970).

Der Chefdesigner bei NSU, Claus Luthe, sah auch nichts Ungewöhnliches darin, das Corvair-Konzept zu übernehmen. Zu attraktiv war diese Designlösung gerade für kleine, kompakt gehaltene Autos, wie es der NSU Prinz 4 von 1964 war. Das Package bot einen vergleichsweise großen Kofferraum im kastenförmigen Bug, im Heck lag der Motor praktisch über der Hinterachse und vermied so das schlechte Kurvenverhalten des Corvair. Und das überstehende Dach war ideal für eine ausreichende Kopffreiheit im Fond, weil es bis zum Abschluß horizontal geführt wurde. Was wollte der Kunde in dieser Zeit mit kleinem Budget mehr als amerikanische Inspiration und deutsche Praktikabilität? Der Werbeslogan traf es haargenau: »Fahre Prinz und Du bist König«. Um wieviel besser mußte er sich fühlen, wenn kurz darauf der gestreckte Prinz 1000, auch in der heißen Version TTS mit hochgestellter Heckklappe zwecks Entlüftung des Motorraums, zu kaufen war und ab 1965 sogar in der Version 110/1200 auf satte 4 m Länge kam. Dabei störte die übertriebene Aluminiumblechblende, die vorgab, ein Grill zu sein.

Wäre der Mazda Familia 1964 mit einem Heckmotor ausgestattet, dann hätte er sich vom Prinz 4 nur durch den fehlenden Dachüberstand unterschieden. So zeigte sein Grill eine für den Frontmotor vergleichsweise großzügige Auslegung in verchromter Manier, dazu die völlig ungestörte umlaufende Chromlinie à la Corvair. Selbst in der Umrißlinie der Radausschnitte glich er

286. Der Chevrolet Corvair zeigt 1960 eine Form, die wohl am häufigsten kopiert wurde: sinnvoll, praktisch und ziemlich einfach zu bauen. (Photo: GM.)

287. 1964 konnte der deutsche NSU Prinz IV ein König sein, fuhr er doch das US-Auto in Miniformat. (Photo: Kunkel.)

288. 1964 ließ der Mazda Familia die horizontale Linie am Bug minimal einknicken. Überflüssig. (Photo: Mazda.)

289. 1965 bot der Suzuki 800 Fronte mit dem Zweitürer speziell am Heck fast schon dauerhafte Eleganz. (Photo: Suzuki.)

290. Fiat hatte 1961 mit dem 1300 das Meisterstück einer Kopie abgeliefert. Dachüberstand, Türgriffe auf Höhe der umlaufenden Chromlinie, glatte Flächen machten das Auto fast zeitlos. (Photo: Fiat.)

286. The design of the 1960 Chevrolet Corvair was probably copied most frequently: sensible, practical and fairly simple to build. (Photo: GM.)
287. In 1964 the German NSU Prinz IV could be a king: It was an American car in miniature. (Photo: Kunkel.)
288. The 1964 Mazda Familia: The horizontal line dips slightly in front. Superfluous. (Photo: Mazda.)
289. The 1965 two-door Suzuki 800 Fronte: close to timeless elegance, particularly in the rear section. (Photo: Suzuki.)
290. In 1961 Fiat produced a masterpiece of a copy, its 1300. Roof overhang, door handles at the level of the chrome swageline, smooth surfaces made the car almost timeless. (Photo: Fiat.)

7.7. The technology is different, the form is modeled on an American one

When tailfins and panoramic windows were at their peak, in 1960, GM sent out on the streets a car that by GM standards was compact. Its own brand, Chevrolet, built the Corvair, a strictly disciplined design. A classic front windshield, curved but not panoramic, a rear window like that of big GM models, a positive C-pillar and roof projection similar to that of the less elegant version of the German Opel Kapitän a year earlier. And here was the characteristic feature of the new design: The completely circumferential beltline was framed in chrome, slightly lowered in front between the headlights in order to replace the missing grille. The undercarriage looked like a bathtub. Making up for the missing grille was the logo on chrome strips and thus pointed to the rear-mounted engine, which needed only a few slits in the hood in order to breathe. No one had even suspected that in technical terms this design was a loser due to its inadequate chassis and heavy tail; nevertheless, it became one of the most copied models, often in Europe, more rarely in Japan. All of the cars had in common the circumferential accent line with its functional advantage: The entire undercarriage was forced into a disciplined volume. And thanks to slender roof pillars the body offered a panoramic view unrivalled by that of any modern car – the excuse being that passive safety today requires substantially stronger pillars and in part has to accommodate airbags, but usually also has to cope with the bright ideas of designers.

There was no fear of that when the 1961 Fiat 1300/1500, the first to be modeled on the Corvair lines, came on the scene. Almost totally it adopted the accent line, which slightly dipped in the rear, mirroring the front. Double headlamps and an identically shaped rear window completed the impression that the car was modeled on the Corvair; and the Fiat, because of the clear configuration of the profile, also represented the trapezoid design – a style that Pininfarina in particular used to employ in the commissioned work he did for Lancia and Peugeot during this period. An inconspicuous design detail of the Fiat were the door handles, which were so slim that they were concealed by the all-around chrome strips. Not a direct copy, but inspired by the Corvair was the BMW 1500, designed by BMW together with the Italian designer Giovanni Michelotti. The vehicle body was classic without the panoramic rear window and roof overhang. The all-around chrome strip was limited to the rear section. On the other hand, the all-around horizontal line was used as a swage line and as a separating groove for the front and rear hood; the door handles were placed in the swage. Together all these details created the impression of a very closed design. And here, for the first time, appeared the so-called Hofmeister kink: the little negative indentation on the rear side window just above the beltline. At the time Wilhelm Hofmeister was the chief designer at BMW (from 1955 until 1970).

The NSU's chief designer, Claus Luthe, also did not think it unusual for his firm to adopt the Corvair concept. This design solution was far too attractive especially for small, compact cars like the 1964 NSU Prinz 4. It had a comparatively large luggage compartment in the box-shaped front, and in the rear the engine was located practically above the rear axle, and thus avoided the

dem NSU, was formal logisch war wegen der generell linearen Auslegung aller bildprägenden Elemente des Corvair-Konzepts. In Details von diesem Konzept gelöst hatte sich der Suzuki Fronte 800 von 1965. Er verzichtete am Bug auf die durchlaufende Linie, um den Scheinwerfern Raum zu geben. Als Ersatz bot er im Heck die runden Doppelhecklichter, so wie sie beim US-Original zu finden waren, hier dank größerer Wagenbreite großzügiger voneinander abgelöst. An der Flanke zeigten die hinteren Radauschnitte ein übliches Motiv der daraus wachsenden, nach hinten gezogenen Sicke. Das Dach fand – ohne Überstand – Abschluß in einem großzügig geformten Heckfenster, das etwas in die Seite geführt wurde und somit auch zur Rundumsicht beitrug. Der Panhard 24 BT von 1965 nutzte das Formenkonzept ab den vorderen Rädern. Der Bug hatte sich am Citroën ID/DS ein Vorbild genommen: flache, gerundete Haube zwischen ovalen, breiten Lichtern. Die ab den Vorderrädern umlaufende Linie wurde dann störungsfrei um das Heck gelegt und fand ihr Pendant in der im Grundriß identischen Linienführung der umlaufenden Regenrinne. Und weil der Panhard eine coupéähnliche Leichtigkeit verfolgte, hatten sich die Hecklichter schmal direkt unter die Corvair-Linie gelegt. Was an diesem Auto, wie auch bei einigen Vorgängern, geschätzt wurde: Sie waren sehr windschlüpfig und konnten mit kleinen Hubräumen flott vorankommen.

Was bei dem ersten Mittelklasse-BMW erfolgreich verkauft wurde, konnte beim BMW 1600 von 1965, dem Vorläufer des 3er-Modells, nicht verkehrt sein. Die Corvair-Linie wurde diesmal in aller Konsequenz umlaufend verchromt bestückt. Das konnte die Trennfugen von Motor- und Heckklappe noch besser verbergen, so wie es heute ohne Chrom in Perfektion des Unsichtbaren viele Modelle von Audi und Skoda schaffen. Die Türgriffe kamen leider nicht beim 1600 auf Höhe der Chromlinie, unbekannt bleibt, ob das konstruktiv bedingt war. BMW wird es wissen. Wir wissen, daß der ZAZ 968 von 1972 eine Doppelkopie ist: vom Corvair und vom NSU Prinz 1000. Zum Unterscheiden packt er sich Frischlufttüten für den Heckmotor an die Flanken über den Hinterrädern. Und 1975 verdoppelt der Hillman Imp Super die Corvair-Linie als schmale Blechfaltung, eine Variante ohne formalen Gewinn. Wir lernen daraus: Das Original ist meist das Bessere.

7.8. Die Technik freut sich über die Form

Bullaugen sind eine kluge Erfindung im Schiffsbau. Sie können dank ihrer Form besser dicht abschließen als ein rechteckiges Fenster. Der Druck auf Dichtungsprofile erfolgt gleichmäßiger, der Einbau ist unkomplizierter, die Funktion ist immer gegeben. Ein gestalterischer Vorteil: Die Kreisform mit einem Mittelpunkt ist richtungslos, kennt keine Achsen wie die Ellipse, die mit der längeren Achse die Richtung vorgibt, liegend, schräg oder senkrecht. Das Bullauge liegt immer richtig und ist leichter in die Umgebung einzubetten als das Rechteckformat eines Fensters. Diese Erkenntnis hilft bei modernen, unruhigen Karosserieseiten, wenn der Tankdeckel seinen Platz zwischen Fugen, Sicken oder Rücklichtgläsern sucht. Das Bullauge ist beim Automobil nicht die richtige Lösung, wenn funktional notwendige Sichtverhältnisse gefragt sind, so bei den Front- und Seitenscheiben. Die Heckscheibe kann weggelassen werden, wenn serienmäßig zwei Außenspiegel montiert sind. Für ein Bullauge bleiben als Fensterfunktion nebensächliche Ausblicke. Die fallen dann auf, wenn alle anderen Fenster große Glasflächen haben und im starken Kontrast zu einem Bullauge stehen.

Der wohl erste Entwurf mit Bullaugen als vollwertige Fenster war 1913 der Alfa von Graf Marco Ricotti. Hier sind sie für die Seitenfenster vorgesehen. Die Sicht nach vorne und schräg in die Seiten verschaffte sich der Fahrer dank der riesigen Panoramascheibe. Beide Fensterformen harmonierten mit dem hintersten Fenster, einer Mischung aus Dreieck und Oval, sowie mit dem zeppe-

291. BMW kann 1962 beim 1500 Front- und Heckdeckelfugen in der Horizontalen verstecken. Die Corvairlinie kann hilfreich sein. (Photo: BMW.)
292. 1965 begnügt sich Panhard bei Seiten und Heck, um die Corvair-Idee zu nutzen. (Photo: Topworldauto.)
293. BMW bleibt 1965 beim 1600 der Corvairlinie treu und läßt die Niere noch unbedeutend erscheinen. (Photo: BMW.)
294. Der russische ZAZ 968 ist 1972 die Kopie der Kopie: erst Corvair, dann NSU Prinz. Ausnahme: die seitliche Hutze. (Photo: ZAZ.)
295. 1975 macht dann auch Hillman mit dem Imp Super das Kopieren mit, aber englisch unbeholfen. (Photo: Simoncars.)

291. In the 1962 BMW 1500, front and rear opening cutlines are concealed in the horizontal line. The Corvair Line can be helpful. (Photo: BMW.)
292. In the design of its sides and rear section, the 1965 Panhard takes advantage of the Corvair idea. (Photo: Topworldauto.)
293. In its 1965 model 1600, BMW remains true to the Corvair Line; the kidney look still has no importance. (Photo: BMW.)
294. The Russian ZAZ 968 (1972) is the copy of a copy: first the Corvair, then the NSU Prinz. An exception is the air scoop on the sides. (Photo: ZAZ.)
295. Hillman with its 1975 Imp Super also copies other designs, though in a heavy-handed British way. (Photo: Simoncars.)

poor handling of the Corvair on curves. And the overhanging roof was ideal for sufficient headroom in the back because it was horizontal right to the end. What more could a customer who had a small budget want? Here was an American-inspired car with German practicability. The advertising slogan hit the nail on the head: »Drive a Prinz and you'll be king«. And how much better you would feel when shortly thereafter the elongated Prinz 1000, including one in the hot TTS version with a raised tailgate (to ventilate the engine compartment), was available for sale, and after 1965, in the 110/1200 version, was even all of 4 m long. At the same time, it had an annoyingly exaggerated aluminum trim panel that professed to be a grille.

If the 1964 Mazda Familia had had a rear-mounted engine, the only difference between it and the Prinz 4 would have been the overhang of the roof. For instance, the dimensions of its chrome-plated grille were comparatively generous for a front-mounted engine, plus it had a completely unbroken chrome accent line à la Corvair. Even the contour of its wheel cutouts resembled that of the NSU, which was logical from a formal point of view because of the generally linear interpretation of all the elements that contributed to the Corvair's image. In certain details, the 1965 Suzuki Fronte 800 diverged from this concept. It lacked the accent line in the front section, instead providing space for the headlamps. To make up for this, there were round double taillights, similar to those of the U.S. original – somewhat farther apart here because the car was wider. At the flank the rear wheel cutouts showed a common motif – a swage line growing out of them and sloping backwards. The roof – without an overhang – ended in a generously proportioned rear window that extended slightly into the side and thus contributed to all-round visibility. The 1965 Panhard 24 BT used the same design concept, starting with the front wheels. Its nose was modeled on that of the Citroën ID/DS: a flat, rounded hood between oval, wide headlamps. The accent line that began near the front wheels then ran unobstructed around the rear section and was matched by the identically laid-out lines of the rain groove that circumnavigated the body. And because the Panhard aimed to achieve the lightness of a coupé, the taillights were placed directly under the Corvair line. Something that people appreciated about this car, as well as about some of its predecessors, was that they were very streamlined and could make good time with small-capacity engines.

There was no reason why features that sold successfully in the first mid-range BMW should not do just as well in the 1965 BMW 1600, the precursor of the 3-series model. This time the Corvair line was very consistently chrome-plated all around. This made it possible to conceal the shutlines of the engine compartment and tailgate lid even better, something that today many Audi and Skoda models manage to do perfectly without chrome. Unfortunately in the 1600 model the door handles were not at the same height as the chrome line, and it is not known whether there is a structural reason for this. BMW knows, no doubt. We know that the 1972 ZAZ 968 is a double copy: of the Corvair and of the NSU Prinz 1000. What is different about the ZAZ 968 is that it has air intakes on the flanks above the rear wheels for the rear-mounted engine. And in 1975 the Hillman Imp Super doubles the Corvair line as a narrow fold in the metal panels, a variant that has no formal advantage. What we learn from this is that the original is usually better than the copy.

7.8. Technology delights in form

In shipbuilding, portholes are a smart invention. Thanks to their form they can be closed much more tightly than a rectangular window. The pressure on gasket profiles is more even, installation is less complicated, the function is always a given. One advantage of this design: The circular form with a center is directionless, has no axes unlike the ellipse, which determines the direction with the longer axis – horizontal, diagonal or vertical. The porthole is always in the right position and can be installed more easily than a rectangular window. It is helpful to be aware of this while designing the modern, cluttered sides of the car body, when the fuel door cover must find its place between shutlines, swage lines or taillight covers. In a car, the porthole is not the right solution when it comes to functionally necessary visibility, for instance, front and side windows. The rear window can be omitted if two standard outside mirrors are installed. The porthole's function becomes negligible. Portholes are conspicuous when all the other windows have large glass surfaces in stark contrast to that of a porthole.

Perhaps the first design that used portholes as full-fledged windows was the 1913 Alfa of Count Marco Ricotti. They were intended to be the side windows. The driver had a view of the front and a diagonal view of the sides thanks to the huge wrap-around front windshield. Both win-

linähnlichen Aufbau. Wie das auf ein damaliges Auge wirkte, ist nicht überliefert; es dürfte mehr als Verblüffung hervorgerufen haben, denn kaum ein Flugobjekt, kaum ein Landfahrzeug nutzte diese Fensterform, um die Sehweise für automobile Bullaugen trainieren zu können. Wenn beim Auto ein Bullauge hinter einem rechteckig geformten Fenster lag, hatte es verständlicherweise einen kleineren Glasanteil und damit reduzierten Ausblick – oder eben, was wahrscheinlich wichtiger war, einen reduzierten Blick nach innen. Das war gewünscht, wenn sich die Karosse herrschaftlich gab mit dem Ziel, daß der Passagier nicht gesehen werden wollte. Heute machen das Vorhänge, verhindern aber damit auch den Ausblick. Diese runden Fenster, öfter kleiner im Durchmesser als die Höhe der übrigen Fenster, nannten sich »opera windows«. Das Beispiel des Szawe Joswin von 1922 bevorzugte ein großes, liegendes Oval und hatte deshalb wohl zusätzlich einen Vorhang dafür und sogar für die rechteckigen Fenster.

Ein anderes Beispiel für automobile Bullaugen nutzte der Voisin C25 Aérodyne von 1934. Vier Bullaugen in Reihe, mittig eingelassen ins Dach, waren vermutlich das erste Beispiel einer Autodachverglasung. Die Steigerung lag in der Möglichkeit, das Dach auf Schienen nach hinten herunterzufahren, weil die Dachneigung gleichförmig rund verlief. Etwas gerader geführt, kam der Porsche Targa 1995 nach Jahrzehnten mit vollflächigem Glasdach. Wer dachte da noch an den Voisin? Nutzt man ein Bullauge ohne Glas, ist es das klassische Loch. Wertet man dies, mit Chrom eingefaßt, auf, so wird es zu einer denkbaren Motorraumentlüftung. Einbau und Lage sind, wie weiter oben beschrieben, problemlos und weniger aufwendig als Schlitze oder Klappen. Eines der ersten Nachkriegsbeispiele ist der Cisitalia 200 von 1947. Hier sind liegende Ovale auf der Flanke des vorderen Kotflügels perfekt in der Fläche plaziert, auf exakt der gleichen horizontalen Linie wie der schlanke, versenkte Türgriff. Diese Qualität durfte erwartet werden, denn der Karosserieentwurf stammte von Pininfarina. Geringere Talente schufen Karossen wie den Mathis 666. Seine Fließheck-Seitenansicht mit Ansätzen eines Heckfensters à la Studebaker Champion Starlight waren handwerklich unbeholfen, nichts fügte sich harmonisch, Türgriffe hingen müde an den Türen, und das Bullaugenmotiv konnte nichts retten. Drei kleine Ringe mit verschließbaren Chromkappen gaben diesem Auto von 1947 so wenig Chic wie die erdige Zweifarbenlackierung. Der Fiat 750 MM Topolino von Zagato legt zwei ovale Bullaugen an die Chromleine und läßt das Motiv vergessen wegen seiner riesigen Seitenfenster, die sich in die Dachebene biegen, als wäre es ein Besichtigungsfahrzeug. Bullaugen mit absoluter Daseinsberechtigung hatte der französische Karosseriebauer Chapron 1950 einmal auf einem Packard Super Eight Chassis von 1937 und einmal auf einem Hotchkiss-Fahrgestell gebaut – als Ausflugdampferchen, vermutlich als Eyecatcher für Werbezwecke.

Bullaugen können schnell zum modischen Attribut werden, das Möglichkeiten der Anwendung probiert. Beim flachen Talbot Lago T26 mit Kleid von Saoutchik reduzierten sich 1950 die Bullaugen auf schmale Schlitze, integriert in Chromleisten, sowohl an der Flanke des vorderen Kotflügels als auch auf der Abdeckung des Hinterrads. Dann lieber Kreis und Bullauge retten, die 1951 dreifach mit Chromringen die Seite des Abarth 205a Berlinetta von Vignale belebten und die Funktion betonten. Dieser Gestaltungsmaxime folgte der Maserati A6G 2000 Spider von 1951 mit dem Design von Pietro Frua. Bullaugen aber hatten ihre größten Karriere 1963 bei Buick. Das Konzeptauto Wildcat II von 1954 legte drei ovale Augen zwischen Haube und Seitenblech und

296. Der Fiat 1100 mit der Karosserie von Coriasco macht 1953 alles richtig: Bullauge zum Schiffsauto. (Photo: Miedema.)
297. 1914 kam der Alfa 40-60HP Ricotti, gebaut von Castagna. Keiner hatte Bullaugen konsequenter angewendet. (Photo: Archiv Autor.)
298. Aus dem Kreis wird 1922 die Ellipse, und schon wirkt der Szawe Joswin vornehm. (Photo: P. Bovyn.)
299. 1934 ist der Blick zum Himmel beim Voisin C25 Aerodyne kein Problem. (Photo: Dan Saranga.)
300. Jedes Talent zur Gestaltung fehlte 1947: Eher wie aus Kinderhand wirkte der Mathis 666. (Photo: Fangio 678.)
301. 1956 kann ein kleines Bullauge elegant wirken, wenn es vom Ford Thunderbird gefahren wird. (Photo: American Cars for Sale.)
302. Es sind keine Gucklöcher, sondern die Luftzufuhr für rasante Motorisierung des Messerschmitt FMR 500 Tiger von 1958. (Photo: Hemmings.)
303. 1950 verwehen Kreise zu Schlitzen beim Talbot Lago T26 mit Saoutchik-Karosse. (Photo: Coachbuild.)
304. Der Buick Wildcat II verschiebt 1954 die Bullaugen als Ellipsen in die Haube – bei einer Studie erlaubt. (Photo: GM.)

296. The 1953 Fiat 1100, with bodywork by Coriasco, gets everything right: a porthole for a boat
car. (Photo: Miedema.)

297. The 1914 Alfa 40-60HP Ricotti, built by Castagna. No other manufacturer used portholes more
consistently. (Photo: author's archive.)

298. In 1922 the circle becomes an ellipse, and
now the Szawe Joswin looks elegant. (Photo: P.
Bovyn.)

299. In a 1934 Voisin C25 Aerodyne, looking at
the sky isn't a problem. (Photo: Dan Saranga.)

300. Like something drawn by a child: The 1947
Mathis 666 shows lack of designer talent. (Photo:
Fangio 678.)

301. In a 1956 Ford Thunderbird, a small porthole
looks elegant. (Photo: American Cars for Sale.)

302. Those aren't peepholes. They're the air supply vents of the 1958 Messerschmitt FMR 500 Tiger.
(Photo: Hemmings.)

303. In 1950 circles have become slits in the Talbot Lago T26 with Saoutchik bodywork. (Photo:
Coachbuild.)

304. The 1954 Buick Wildcat II shifts the portholes – now ellipse-shaped – to the hood, which
is permissible in a study car. (Photo: GM.)

dow forms harmonized with the rearmost window, a mixture of triangle and oval, and with the
Zeppelin-like body. We do not know what effect this had on people at the time; it must have provoked more than amazement, for hardly any aircraft or land based vehicle used this type of window, and people were not used to seeing portholes on cars. A porthole on a car, behind a rectangular window, understandably had a smaller pane and thus limited the view of the outside – or
simply, which was probably more important, it was more difficult to look inside. That was a desir-
able feature if the ladies and gentlemen inside the car did not want to be seen. Today this is done
by means of curtains, but curtains also keep you from looking out of the window. These round
windows, often with a diameter smaller than the height of the other windows, were called »opera
windows«. One car, the 1922 Szawe Joswin, preferred a large horizontal oval, which is probably
why it had an additional curtain for the porthole and even for the rectangular windows.

Another car that used portholes was the 1934 Voisin C25 Aérodyne. Four portholes in a row,
recessed in the center of the roof, were presumably the first example of automotive roof glazing.
An added feature was the fact that the sunroof could be made to slide down backwards on
tracks, because the slope of the roof was uniformly round. Somewhat more straight-lined,
decades later, the 1995 Porsche Targa came up with a roof made completely of glass. Who still
remembered the Voisin? If a porthole without glass is used, it is a proverbial hole. If it is given
added value by framing it in chrome, it conceivably turns into engine-compartment ventilation.
Its installation and positioning, described above, are hassle-free and less expensive than slits or
flaps. One of the first postwar examples is the 1947 Cisitalia 200. Here horizontal ovals on the
flank of the front fender are perfectly placed in the panel, on exactly the same horizontal line as
the slim, recessed door handle. This quality was only to be expected, for the bodywork design
was by Pininfarina. Lesser talents created bodywork like that of the Mathis 666. Its fastback side
profile with what looked vaguely like a rear window à la Studebaker Champion Starlight were
awkwardly constructed, nothing fit harmoniously, door handles drooped on the doors, and the
porthole motif could not save the situation. Three small rings with closable chrome covers gave
this 1947 car as little chic as did its earthy two-tone paintwork. On the Fiat 750 MM Topolino by
Zagato, two oval portholes are placed on the chrome line but we forget the motif because of the
car's huge side windows, which curve up into the roof as though this was a sightseeing bus. In
1950, the French car body manufacturer Chapron built portholes that had an absolute right to ex-

eröffnete damit, neben den schnellen Linien an den Flanken, das neue Buick-Wiedererkennungs-motiv. Ein Jahr später wurden die drei Buick-Augen für den Buick Special reserviert, die auch das Roadmaster-Modell zieren. Sie durften, zu Schlitzen mutiert, 2007 noch einmal beim Buick Riviera Konzeptauto wiederkehren.

Bei Ford hat man das Bullauge nicht für die Motorraumentlüftung verwendet. Die Serienversion des Ford Thunderbird hielt für das Hardtop ein kleines Opera-Window parat und gab 1956 dem Modell sportliche Vornehmheit, nur drei Kreise belebten die ruhige Seitenansicht, zwei Räder und das Opera-Window, was der breiten C-Säule die Wucht nahm. Bullaugen waren für Buick in den 1950er Jahren das Wiedererkennungsmerkmal. Bei dem deutschen Sportwagen Borgward Hansa 1500 RS von 1958 waren sie das nur der Funktion zugeschriebene Detail der Motorraument-lüftung, haben aber immer wieder wegen der eindeutigen Kreisform und der linearen Anordnung eine ornamentale Wirkung gehabt und dem automobilaffinen Auge Gedanken an Buicks aufkom-men lassen. Wenn Bullaugen für das sportlich ausgelegte Auto eine sinnvolle, weil eine praktische Lösung sind, um dem Motorraum Frischluft zu gönnen, dann dürfte auch der kleinste Vertreter der Sportler dieses Rezept verwenden. Wir meinen den vierrädrigen Messerschmitt FMR 500 Ti-ger von 1958. Abgeleitet von der dreirädrigen Variante mit erstarktem Motor, war dieses Wägel-chen mit 19,5 PS der Schrecken mancher Mittelklasse, wenn es mit Topspeed von 126 km/h an den Großen vorbeizog. Je zwei Bullaugenpärchen pro Seite belüfteten den Heckmotor. Seine ins-gesamt 320 gebauten Exemplare lassen den Kaufpreis nach sechs Jahrzehnten leicht sechsstel-lig geraten.

Als modisches Attribut sind die Bullaugen ohne jegliche Funktion, es sei denn, Schmuckwerk hat eine psychologische Aufgabe. Nicht anders ist das Beispiel des Fiat 125s Samantha Coupés zu sehen, das Vignale 1968 baute. Hier hätte das gelungene Fließheck auf Bullaugen in Kleinstfor-mat durchaus verzichten können. Sie waren zu klein, um als Gucklöcher ausgebildet zu werden, was Sinn gehabt hätte bei der großen, geschlossenen Fläche der C-Säule. Oder man übertreibt das Thema, läßt die Zahl von Bullaugen inflationär anwachsen und schafft damit fast eine Loch-struktur wie beim Aston Martin DBS Sotheby Special von Ogle Design von 1972. Hier zeigte sich die Heckwand als Lochwand, teilweise belegt mit den Hecklichtern. Daß solche Orgien von kreis-förmigen Löchern im Wageninneren – Bullauge dafür als Begriff zu wählen, wäre fast unange-bracht – Designer herausfordern, konnte mit dem Lancia Beta Trevi 1980 nachgewiesen werden. Hier trugen unterschiedlich große, kreisrunde Vertiefungen alles, was beim Armaturenbrett zu fin-den ist: Tacho, Drehzahlmesser, Öldruckanzeige, Uhr, Benzinanzeige und viele, viele Knöpfe, alles in die Löcher eingegraben.

Zurück zu den wahren Bullaugen, denn die sind beim Plymouth Expresso ihrem Namen ange-messen. Sie werden hier dem ursprünglichen Sinn, der ursprünglichen Funktion nach richtig ge-nutzt, sie bilden die Seitenfenster. Und weil sie so groß sind, wirken sie eher seltsam als funktio-nal, was vielleicht an unserer Sehweise liegt, die Bullaugen so konnotiert, daß es kleine Glasfens-ter auf großen, geschlossenen Flächen seien. Das änderte sich auch nicht, als 1994 der Expres-so als Concept-Car gezeigt wurde. Der Chrysler-Konzern, zu dem auch Plymouth gehörte, nutzte das Bullaugenmotiv 1998 bei einem weiteren Concept-Car, dem Chrysler Chronos. Er hatte (wuß-ten das seine Designer?) wie der Fiat 750 MM auf jeder Seite am oberen Flankenbereich zwi-schen Vorderrad und Türöffnung je zwei elliptische Bullaugen, die auf einer zierlichen Linie aufge-fädelt wurden, um die Motorraumlüftung zu fördern. Kennt man die Gestaltungsabsichten von Designern, dann war, abgesehen von der Funktion der Motorbelüftung, zumindest auch eine symbolische Bedeutung für ein sportlich ausgelegtes Fahrzeug vorgesehen. Daß es beim Allard J2X MK I von 2006 um das rein Funktionale ging, darf vermutet werden, denn nichts an diesem

305. 1955 und auch sonst: Der Buick, hier als Spe-cial, war für seine drei Bullaugen bekannt. (Photo: Kenora 58.)

306. Kaum auszumachen sind sie 1968, eher Lö-cher im Käse beim Fiat 125s Samantha von Vig-nale. (Photo: Autodato.)

307. 1972 kann man es auch übertreiben am Heck des Aston Martin DBS V8 Sotheby Special. (Photo: Archiv Autor.)

308. Kein kalter Kaffee 1994: Der Plymouth Es-presso übertreibt das Motiv – verzeihlich bei einer Studie. (Photo: Plymouth.)

309. 1998 gefällt das Gesamtdesign der Studie Chrysler Chronos, weil die Bullaugen die Ruhe der Coupélinie bewahren. (Photo: Chrysler.)

310. Der Allard J2X MK1 sieht 2006 älter aus, als er ist, dabei ist das Bullaugenmotiv zeitlos. (Photo: Allard.)

311. 2013 mutieren beim Maserati Quattroporte S9 die Bullaugen zu schrägen Rauten, seit langem hat die Marke sonst runde. (Photo: Maserati.)

312. Der Borgward Hansa RS 1500 kann es 1958 sportlich so gut wie Porsche, inklusive Bullaugen. (Photo: Steenbuck.)

305. In 1955 and other years: Buicks were known for their three portholes. Pictured is a Buick Special. (Photo: Kenora 58.)

306. You can hardly tell they're there: in the 1968 Fiat 125s Samantha by Vignale they look like holes in Swiss cheese. (Photo: Autodato.)

307. Overdone? The rear section of the 1972 Aston Martin DBS V8 Sotheby Special. (Photo: author's archive.)

308. Not old hat at all: The 1994 Plymouth Espresso exaggerates the motif – excusable in a study. (Photo: Plymouth.)

309. The overall design of the 1998 study Chrysler Chronos is pleasing, because the portholes preserve the equilibrium of the coupé line. (Photo: Chrysler.)

310. The 2006 Allard J2X MK1 looks older than it is. At the same time the porthole motif is timeless. (Photo: Allard.)

311. In the 2013 Maserati Quattroporte S9 the portholes mutate into diagonal diamond shapes. For years, the brand has usually had round portholes. (Photo: Maserati.)

312. The 1958 Borgward Hansa RS 1500 is as sporty as the Porsche, including the portholes. (Photo: Steenbuck.)

ist, once on a 1937 Packard Super Eight chassis and once on a Hotchkiss chassis – as little excursion steamers, presumably as eye-catchers for advertising purposes.

Portholes can quickly become a fashionable attribute for trying out a variety of possible uses. In the low-slung 1950 Talbot Lago T26 with a body by Saoutchik, the portholes were reduced to narrow slits integrated in chrome strips both on the flank of the front fender and on the cover of the rear wheel. Then why not salvage the circle and the porthole motifs? In 1951, the designer used three portholes surrounded by chrome in order to jazz up the side of the Abarth 205 a Berlinetta by Vignate and to emphasize its function. The 1951 Maserati A6G 2000 Spider with a design by Pietro Frua followed this maxim. But portholes played their most important role in the 1963 Buick models. In the 1954 concept car Wildcat II, three oval eyes were placed between the hood and the side panel, marking the beginning of a new Buick recognition factor, in addition to the fast lines on the flanks. One year later the three Buick eyes were reserved for the Buick Special, and they also grace the Roadmaster model. Mutated to slits, they would show up once again in the 2007 Buick Riviera concept car.

Ford did not use the porthole for engine-compartment ventilation. The production version of the Ford Thunderbird had a small opera window; in 1956 the model was given sporty elegance: only three circles enlivened the placid profile, two wheels and the opera window, which reduced the impact of the C-pillar. In the 1950s portholes were the Buick hallmark. In the 1958 German sports car Borgward Hansa1 500 RS, they were there for engine-compartment ventilation, solely functional, but they always had an ornamental effect due to their unmistakable circular shape and linear arrangement, and the fact that they always reminded car-conscious eyes of Buicks. If portholes, because they are practical, are a sensible solution for a sports car, since they supply air to cool the engine compartment, then even the smallest sports car might be expected to use this recipe. We mean the 1958 four-wheel Messerschmitt FMR 500 Tiger. Derived from the three-wheel variant that had a stronger engine, this little car, at 19.5 hp, was the terror of many a mid-size car when it overtook the big guys at a top speed of 126 km/h. Two little pairs of portholes per side ventilated the rear-mounted engine. Since a total of only 320 copies were ever built, its price six decades later could easily be in the six-figure range.

As a fashionable attribute portholes have no function whatever, unless there's a psychological reason behind decorative details. There's no other explanation for the Fiat 125s Samantha coupés that Vignale built in 1968. The successful hatchback could have perfectly well done without its mini-sized portholes. They were too small to be peepholes, which would have made sense in the big, closed panel of the C-pillar. Or else the theme is exaggerated, with the number of portholes growing at an inflationary rate, thus almost creating a perforated structure as in the 1972 Aston Martin DBS Sotheby Special by Ogle Design. Here the back panel was a wall of perforations, in part incorporating the taillights. As the 1980 Lancia Beta Trevi demonstrates, such orgies of circular holes in the car's interior are a challenge to designers – portholes are not the only example. In this model, large, circular depressions of varying sizes housed everything that is found on a dashboard: speedometer, r.p.m. counter, oil pressure gauge, clock, fuel gauge and many, many buttons, all recessed in the holes.

But back to the true portholes, for in the Plymouth Expresso these appropriately reflect their name. Here they are correctly used in their original sense, their primary function: They are the side windows. And because they are so large, they tend to look strange rather than functional, which

Auto ist auf gestalterischen Effekt ausgelegt. Seine Formensprache vermeidet jeden Versuch, sich modern und dynamisch zu geben. Das erwartet man von einem Engländer, der konservative Sportlichkeit bevorzugt.

Ganz anders die Italiener: Sie hatten in allen Jahrzehnten seit den 1950er Jahren bewiesen, daß ihre Talente automobiler Gestaltung nur selten für Modisches zugänglich waren. Maserati hatte zum Beispiel das Bullaugenmotiv entsprechend dezent in einer Häufigkeit verwendet wie einst Buick und oft auf schmale, rechteckig anmutende Luftschlitze reduziert – oder wie beim Quattroporte S9 von 2013 als rhombenähnliche Trios an klassischer Stelle der Flanken. Zu vermuten ist: Bullaugen werden auch zukünftig zu finden sein.

7.9. Form und Technik glänzen am Heck

Gewöhnlich hat das Auto zwei Rücklichter, zwei Bremslichter und später das Rückfahrlicht in Weiß sowie rotes Nebelrücklicht. Das dritte Bremslicht hatte hier nichts zu suchen und verabschiedete sich auf die Kante am Stufenheck oder unter das Dach. Bleibt die Leuchtenkombination: Erste zaghafte Chromstreifen schufen die Verbindung von linker und rechter Lichtanlage, was die Zusammengehörigkeit unterstrich und der Heckansicht etwas optische Breite suggerierte. Der Chromstreifen teilte sich manchmal auf und bildete so den Rahmen um das Lichterglas, eine weitere Aufwertung. Ein Beispiel dieser Zusammenarbeit von Lichttechnik und Form war der Fiat 1300 mit dem Kleid des Karossiers Moretti. Weil es 1963 war, hatten die Lichter noch Glühbirnen und zur besseren Lichtausbeute kleine Hohlspiegel. Das wiederum bedurfte eines nicht zu kleinen Bauteils, um seine Funktion zu erhalten. Beim Moretti hatte der kräftige Chromsteg zwischen den Rücklichtern die einfache ovale Leuchtenform aufgewertet. Weil die runden Rückfahrlichter mit dem Klarglas gesondert daruntergelegt wurden, gab es farblich keine Diskrepanz, als wären sie mit den roten Lichtern verbunden. So macht die komplette Heckansicht einen aufgeräumten Eindruck, weil ganz unten die hörnerfreie Stoßstange mit dem kräftigen Chromsteg harmonierte. Dieses Rezept kann auch dem Mercedes SL von 1963 zugeschrieben werden, hier in der restaurierten Ausgabe aus dem Hause Brabus von 2015. Im Gegensatz zum Moretti mit hohem Heckabschluß hatte der SL einen flachen Abschluß, chromgefaßte Hecklichter rückten dicht an die geteilte Stoßstange. Der verbindende Chromsteg blieb als Rahmen der Heckleuchten profilidentisch und wirkte daher insgesamt zierlicher. Beide Beispiele waren Vorboten zur heutigen Tendenz intensiven Chromschmucks für die Hecklichter.

Der Weg dahin hatte mit dem Lancia Trevi von 1982 den Ansatz einer Steigerung gefunden. Die Umsetzung aber war nur bescheidene Gestaltung. Das gewöhnliche Stufenheck mit glattem,

313. 1963 war alles bescheidener, auch Chrom am Heck, hier bei der aufgearbeiteten Version des Mercedes 280 SL 2015. (Photo: Brabus.)

314. Der Fiat 1300 von Moretti ist 1963 ebenso bescheiden wie ein SL aus demselben Jahr. (Photo: Autobelle.)

315. 1982 ist anstelle von Chrom eine matt polierte Aluleiste Namensträger für den Lancia Trevi. Das kannte ein Renault R 10 bereits 1968. (Photo: var-auto.)

316. Ein kräftiger Chromstreifen veredelt 2000 das Heck des Opel Vectra RS, Buchstaben aufwendig erhaben appliziert. (Photo: Opel.)

317. 2006 zieht das Ford Focus Cabrio nach, preiswerter mit gestanzten Buchstaben. (Photo: Ford.)

318. Der Saab 9-5 macht 2009 eine Variante aus Glas und mattem Alu, zurückhaltend und fast edel. (Photo: Saab.)

313. In 1963 everything was less pretentious, including chrome in the rear section. Pictured here is the refurbished version of the Mercedes 280 SL 2015. (Photo: Brabus.)
314. The 1963 Fiat 1300 by Moretti is just as modest as an SL built the same year. (Photo: Autobelle.)
315. In 1982, instead of chrome, a matt, polished aluminum strip bears the name of the Lancia Trevi. This was already a feature of the 1968 Renault R 10. (Photo: varauto.)
316. A bold chrome strip adds an elegant touch to the rear section of the 2000 Opel Vectra RS; the letters are raised, a costly process. (Photo: Opel.)
317. The 2006 Ford Focus Cabrio follows suit, but has less expensive stamped letters. (Photo: Ford.)
318. The 2009 Saab 9-5 gives us a glass and matte aluminum version, restrained and almost aristocratic. (Photo: Saab.)

may have something to do with the way we perceive things: Our connotation of portholes is that they are small glass windows in large, closed surfaces. That didn't change even when, in 1994, the Expresso was revealed as a concept car. The Chrysler company, which also produced the Plymouth, used the porthole motif in 1998 in yet another concept car, the Chrysler Chronos. Like the Fiat 750 MM (did its designers know?) it had two elliptical portholes each in the upper flank area on both sides between the front wheel and the door opening, arranged in a graceful line, their purpose being the ventilation of the engine compartment. Knowing what we know about the designers' design intentions, the portholes were at least intended to have symbolic significance for a sports-oriented vehicle, apart from their function of ventilating the engine compartment. We may assume that in the 2006 Allard J2X MK I, portholes were purely functional, for nothing about this car can be interpreted as a purely design-based effect. Its stylistic idiom makes no attempt to appear modern and dynamic. That is to be expected of a conservative British sports car.

In this respect the Italians are totally different: During all the decades since the 1950s they demonstrated that their talents in automotive design only rarely extended to trendy details. Maserati, for instance, had subtly used the porthole motif with the same frequency as Buick did once, and often reduced it to narrow, rectangular-looking air slots – or, as in the 2013 Quattroporte S9, as lozenge-shaped trios, in the classic position on the flanks. We assume that portholes will be featured in future cars as well.

7.9. Form and technology gleam on the rear

Normally a car has always had two taillights, two brake lights, and later the backup light in white as well as a red rear fog light. The third brake light was out of place here and was relocated to the edge of the notchback or below the roof. Then there's the combination of lights. The first hesitant chrome strips created a link between the left-hand and right-hand light system, underlining the fact that they go together and making the rear view of the car appear somehat wider. The chrome strip sometimes divided, framing the taillight covers, an additional upgrade. One example of the collaboration between light technology and form was the Fiat 1300, with bodywork designed by the car body manufacturer Moretti. Because it was 1963, the lights still had lightbulbs, and small concave mirrors to improve luminous efficacy. This in turn required a sizable component part in order to function. In the Moretti, the hefty chrome bar between the taillights upgraded the simple oval shape of the lights. Because the round backup lights with their clear glass were positioned separately below them, there was no discrepancy as far as colors were concerned, as though

319. Der Opel Vectra macht es 2013 richtig: Erst drückte die Chromleiste auf die Hecklichter, jetzt liegt sie mittendrin. (Photo: Opel.)
320. 2009 könnte diese Variante des Tesla Model S ein Jaguar-Motiv sein. (Photo: Tesla.)
321. Auch ein SUV muß 2012 garniert werden. Die Chromlinie paßt gerade noch ins Glas des Mercedes GL. (Photo: Daimler.)
322. 2013 könnte diese Variante des Jaguar XF ein Tesla-Motiv sein. (Photo: Jaguar.)
323. Der Maserati Quattroporte macht 2013 die Chromleistenmode mit, aber dünn und fein gebogen. (Photo: Maserati.)
324. 2013 zeigt uns die Studie H6 von Pininfarina, wie man Chromleistenmode zeitlos macht. (Photo: Pininfarina.)
325. Der BMW 740e kann 2016 dem Trend nicht widerstehen, verschlankt aber das Chrom im Glasfeld. (Photo: BMW.)

senkrechtem Abschluß und allgegenwärtigen Rechteckleuchten bekam an der Heckdeckel-Unterkante eine Blende aus mattem Aluminium mit Firmenschriftzug und Modellname. Da sie über den Leuchten lag, fehlte die schlüssige Verbindung und wirkte als zusätzlich applizierte Aufhübschung, damals ein neues Motiv. Es dauerte lange, bis Opel mit dem Vectra 2002 zeigte, wie dieses Thema in gute Gestaltung umgesetzt werden kann. Inzwischen wuchsen die Rücklichter. Rotes Glas und Klarglas kombiniert, gaben beim Opel der Idee des die Lichter verbindenden Chrombands eine gestalterische Logik, denn dieses verbindet sich optisch mit dem Klarglas und verlängerte somit die hell erscheinende Unterkante der Heckklappe. Das Detail der aufgelegten Buchstaben auf dem Chromband erhöhte die Wertigkeit, was bei gestanzten und somit vertieften Buchstaben ausbliebe. So machte es das Ford Focus Cabrio von 2006, das ansonsten dasselbe Gestaltungsrezept der verbindenden Chromleiste hatte.

Der Prototyp der aktuellen Designheckleuchten mit Chromspange konnte 2009 beim Tesla Model S studiert werden. Hier wurden die Rücklichtgläser auf die spitz auslaufende Chromspange aufgespießt und auf ewig miteinander vermählt. Der Markenname war sanft in die Chromfläche gestanzt wie beim Focus. Und unauffällig integrierten die Chromspitzen die Rückfahrlichter. Alles ein Beispiel dafür, daß Mode nicht präpotent das Auge strapaziert. Diese Zurückhaltung war auch beim Saab 9-5 zu sehen, der bei seiner Premiere 2009 nur noch wenige Jahre vor sich hatte, weil die Marke 2012 ihr Ende sah. Hier reduzierte sich die Spange zu einem Streifen aus mattem Aluminium, der eine weitere Betonung durch die Acrylglasspange darüber erfuhr. Beides aber wurde nur für den Schriftzug »Saab« genutzt. Ohne die Aufgabe als Funktionsträger für den Marken- oder Typenname putzte der Mercedes GL sein Heck mit einer Chromspange heraus, die Nachweis dafür ist, daß selbst veritable Geländewagen möglichst viele »zivile« Schmuckdetails nutzen, um ihre Wucht zu mildern. Hier legt sich die Spange noch vergleichsweise harmonisch an die Rücklichter. Einige Mercedes-Modelle haben sich hier schwergetan, die formale Verbindung formgerecht zu erreichen: Mal drückte die Spange auf das Lichterglas, mal langte es, nur die Breite der Heckklappe zu nutzen, ohne die Hecklichter zu erreichen – ein Merkmal, als sei es nachträglich appliziert, weil es zur Mode geworden war, Chrom in die Heckansicht zu pflanzen. Der überarbeite Opel Insignia von 2013 verbesserte das zuvor unglücklich plazierte Chromband ähnlich wie bei Mercedes nun in die Heckgläser, die sich dafür winklig geben, als wären sie umgedrehte L-förmige Lichter aus der BMW-Designsprache.

Der Jaguar XF von 2013 ist ein weiteres Beispiel und ein Zwilling der Tesla-Gestaltungsidee von Chromband und Hecklichtern. Der Trend zu inflationären Ideen identischer Formgebung wird

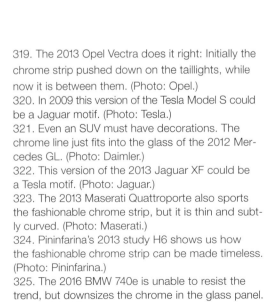

they were connected with the red lights. Thus the total look of the rear gives the impression of being uncluttered, because all the way at the bottom the bumper, which has no guards, matched the sturdy chrome bar. This recipe is also followed by the 1963 Mercedes SL, pictured here in the restored 2015 version by Brabus. In contrast to the Moretti, which has a high rear end, the SL's rear is flat, and chrome-framed taillights have moved closer to the divided bumper. The connecting chrome bar – a frame for the taillights – had remained identical in profile and thus looked far more graceful on the whole. Both examples were precursors of the current trend of having lots of chrome taillight decoration.

It was the 1982 Lancia Trevi that had been heading in that direction. But the Lancia's design was rather modest. Attached to the lower edge of the trunk lid of a standard notchback, with a smooth, vertical rear end and omnipresent rectangular lamps, was a matte aluminum panel bearing the firm's logotype and the name of the model. Since it was located above the taillights, there was no logical connection, and the effect was of an added-on embellishment, then a novel feature. It did not take long for Opel to demonstrate, with its 2002 Vectra, how this theme can be transformed into good design. Meanwhile taillights grew in size. A combination of red glass and clear glass in the Opel showed that the idea of a chrome strip connecting the lights made sense as a design, for the chrome is visually linked with the clear glass, thus making the bright lower edge of the tailgate appear longer. The detail of the letters superimposed on the chrome strip increased the car's intrinsic value, which would not have been the case if the letters were stamped into the metal and thus recessed. That was true of the 2006 Ford Focus Cabrio, which in other respects used the same design recipe, a connecting chrome strip.

The prototype of current designer rear lights plus chrome bar could be studied in the 2009 Tesla Model S. Here the glass covers of the taillights were skewered on the sharply tapering ends of the chrome bar and forever joined together. The brand name was gently stamped into the chrome surface, as in the Focus. And the tips of the chrome bar unobtrusively integrated the backup lights. It all goes to show that fashion does not have to be a strain on the eyes. This restraint could also be observed in the Saab 9-5, which at its premiere in 2009 had only a few years ahead of it, since the brand went out of business in 2012. Here the bar was reduced to a strip of matte aluminum, which was further emphasized by the acrylic-glass bar above it. But both were used only for the logo »Saab«. Since the rear of the Mercedes GL did not have to display the brand or model name, it was spiffed up with a chrome bar, proving that even honest-to-goodness all-terrain vehicles use a plethora of »civilian« ornamental details to soften the full force of their impact. Here the bar sits comparatively harmoniously next to the taillights. It's been difficult for some

319. The 2013 Opel Vectra does it right: Initially the chrome strip pushed down on the taillights, while now it is between them. (Photo: Opel.)
320. In 2009 this version of the Tesla Model S could be a Jaguar motif. (Photo: Tesla.)
321. Even an SUV must have decorations. The chrome line just fits into the glass of the 2012 Mercedes GL. (Photo: Daimler.)
322. This version of the 2013 Jaguar XF could be a Tesla motif. (Photo: Jaguar.)
323. The 2013 Maserati Quattroporte also sports the fashionable chrome strip, but it is thin and subtly curved. (Photo: Maserati.)
324. Pininfarina's 2013 study H6 shows us how the fashionable chrome strip can be made timeless. (Photo: Pininfarina.)
325. The 2016 BMW 740e is unable to resist the trend, but downsizes the chrome in the glass panel. (Photo: BMW.)

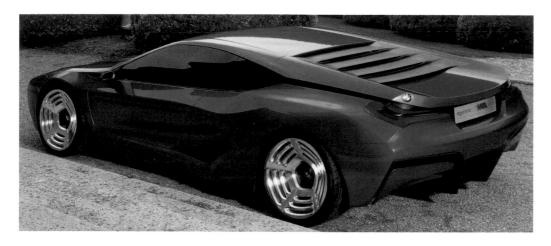

326. 2008 kann die BMW-Studie Turbo nur einen Schlitz zwischen oben und unten bieten, als Sportler baut er eben extrem flach. (Photo: BMW.)
327. Die Studie Lancia Granturismo macht 2003 die Dachtrennung von der C-Säule eleganter als alle anderen. (Photo: Lancia.)
328. 2009 kann der Dacia Duster Concept das Motiv der unterbrochenen C-Säule raffiniert mit Hecklichtern kombinieren. (Photo: Dacia.)
329. 1983 erinnert die geriffelte Partie des Renault Fuego GTX an einen Balg, als könnte der Aufbau höhenverstellbar sein – ist es aber nicht. (Photo: Riley.)

letzten Endes zu eines Entwertung des Wunsches, sich auch am Heck individuell zu geben. Also stellt sich Frage, Alternativen oder eine andere Gestaltung des Modetrends zu finden. Maserati, wie bereits am Beispiel der Bullaugen gezeigt, macht mit im Trend, läßt das Chromband aber nur zierlich zwischen den Lichtern leben, fast reduziert auf den Griff für die Heckklappe. 2013 zeigt das bei diesem Maserati Quattroporte die vornehme italienische Art. Vornehm kann es auch der Aston Martin Lagonda, als er 2015 ein schlankes Heck durch Aushöhlung noch stärker mit einem präzisen Chromband betont, das die volle Wagenbreite nutzt. Beim Hyundai i40 Kombi bleibt es 2015 bei der Standardlösung, ein modisch angesagtes Attribut zu übernehmen. Und der Mazda 6 Kombi versucht im selben Jahr, das Chromband auf die Gläser zu legen, und muß sie deshalb leicht hochbiegen. In der Mitte reitet das Logo auf dem Band, und der Eindruck sagt dem Auge: fast vergessen, mußte es noch schnell seinen Platz finden. Der BMW 740e von 2016 bleibt nicht außen vor, wenn der Klub der Heckchrombänder angesagt ist. Wo sein formales Konzept beim zeitgleichen Audi A8 etwas eleganter, weil zierlicher, ausgeführt ist, da wird der Chromstreifen zwischen den Lichtern kräftiger und geht nur schlank in die Glasflächen. Nur ein Concept-Car kann sich 2017 Filigranes leisten wie der H600 von Pininfarina. So aggressiv sich der Bug mit seinen senkrechten Chromstegen gibt und damit ansatzweise Zähne zeigt, langt hier ein schmales Chromband, das sich links und rechts noch zum Rahmen teilt, für schlankste rote Glasflächen. Wo andere das Chromband mittig in die Heckgläser legen, hier wird das Band zum Chromrahmen und die Innenfläche mattrot ausgelegt. Würde so etwas jemals die allgemeine Betriebserlaubnis erhalten?

7.10. Technik und Form in Schwarz zur optischen Täuschung

Wie dankbar sind Designer für die schwarze Farbe, seitdem diese wie ein Make-up verwendet wird. Ein solches will nicht nur die Schönheit unterstreichen, sondern auch kleine Makel verbergen oder mit geringen Mitteln Größeres vortäuschen. Der Blick zu den Beispielen findet viele Anwendungsvarianten von Schwarz. Noch in logischer Absicht nutzte der Renault Fuego von 1983 die Farbe, um die geriffelte, überbetonte Fuge zwischen Unterbau und Aufbau zu betonen. Einem gefalteten Blasebalg nicht unähnlich, trennt sie die Karosse so, als könnte sie nach oben wachsen. Ein funktionaler Grund ist nur schwer zu erkennen. So bleibt es bei der möglicherweise falschen Interpretation der schwarzen Fuge, die aber dem Fuego auf Dauer ein unverwechselbares Äußeres gegeben hat. Was bei diesem die überbetonte Trennung von Karosseriebereichen war, kann als Vorbild dienen, um es zwischen Dachsäulen und Dach auszuprobieren. Wenn die C-Säule sich dank geschwärzter Fuge vom Dach absetzt, dann beginnt dieses zu schweben. Sehr früh, aber danach kaum besser gestaltet, zeigte sich das Konzeptauto Lancia Granturismo Stilnovo von 2003. Keiner nach ihm schaffte die Klarheit von der Glasfläche des Heckfensters, der Blechfläche des Daches und der Fläche der C-Säule, alle Partien mit leichter Flächenkrümmung. Der BMW M1 Hommage wußte die Fuego-Idee mit der des Lancia zu verbinden, weil seine flache Silhouette 2008 das Trennende von Dach und Unterbau in die Länge ziehen konnte, was die Leichtigkeit der Karosserie beförderte, anstatt über dem Hinterrad eine zusammenhängende Fläche zu bilden, falls diese Trennung fehlen würde. Sein Vorgänger traute sich 1978 hier nur eine langgezogene, schwarz geschuppte Kieme zur Mittelmotorbeatmung, bereitete das Thema aber damit vor.

of the Mercedes models to express the formal connection correctly: At times the bar was pressed against the cover of the lights, at others it was enough to use the width of the tailgate alone without reaching the taillights – a feature that looked as if it had been applied retroactively because it had become the fashion to add chrome in the rear of the vehicle. The reworked 2013 Opel Insignia improved things by putting the chrome strip, which had previously been badly placed, in the taillight covers like in the Mercedes; these now became angled as though they were L-shaped lights borrowed from the stylistic idiom of BMW.

The 2013 Jaguar XF is yet another example, and a twin of Tesla's design concept – the chrome strip and taillight combination. In the long run the inflationary trend of implementing identical design ideas invalidates the wish to also have a distinctive rear section. At this point it is necessary to find alternatives, or a different form for the fashion trend. Maserati, as demonstrated by the example of the portholes, plays along with the trend, but merely positions the chrom strip gracefully between the taillights, almost reduced to a handle for the tailgate. In 2013, in this Maserati Quattroporte, this is an expression of the distinguished Italian style. The 2015 Aston Martin Lagonda, too, can look distinguished when a slender rear section is emphasized even more by recessing and adding a precise chrome strip that uses the entire width of the car. The 2015 Hyundai i40 station wagon sticks to the standard solution of adopting a trendy detail. And that same year the Mazda 6 station wagon tries to put the chrome strip on the taillight covers, and therefore has to bend them up slightly. In the center the logo rides on the strip, and we have the impression that it had almost been forgotten and the designer had to find room for it at the last moment. The 2016 BMW 740e also joins the club when it comes to having a chrome strip on the tailgate. While the formal concept of the Audi A8, built at the same time, is somewhat more elegantly – and gracefully – implemented, the chrome strip between the taillights becomes more robust here and is only slenderly inserted into the glass taillight covers. Only a concept car in 2017 can afford filigree details like those of Pininfarina's H600. No matter how aggressive the front looks with its vertical chrome bars, seeming to bare its teeth, here a narrow chrome strip is sufficient, dividing on the left and right to frame the slimmest of red glass surfaces. While others place the chrome strip in the middle of the taillight covers, the strip here turns into a chrome frame and the inner surface is lined in matte red. Would anything like this ever get a general operating permit?

7.10. Technology and form in black create an optical illusion

How grateful designers have been for the color black ever since it was first used as a kind of makeup. It not only emphasizes beauty but also hides small defects or, with limited means, simulates something greater. Our examples show us many ways black has been used. With logical intent, the 1983 Renault Fuego used the color in order to draw attention to the rippled, overempha-

326. The 2008 BMW study Turbo only has a slit between the upper and lower section – as a sports car it's simply built extremely flat. (Photo: BMW.)
327. The 2003 study Lancia Granturismo does the separation of the roof from the C-pillar more elegantly than all the others. (Photo: Lancia.)
328. The 2009 Dacia Duster Concept: a sophisticated design combining the motif of the interrupted C-pillar with taillights. (Photo: Dacia.)
329. The rippled section of the 1983 Renault Fuego GTX is reminiscent of a bellows, as though the height of the bodywork was adjustable – which it isn't. (Photo: Riley.)

Ausgerechnet die Marke preiswerter Automobile, nämlich Dacia, hatte seinen ersten, völlig neu gedachten Concept-Car 2009 präsentiert und Duster getauft. Hier wird die C-Säule zu zwei Hälften umgeformt, getrennt durch eine breite, schwarzrote Fuge, die tiefer in Richtung Rad liegt als das vom Lancia bekannte Motiv. So bilden die Hälften der C-Säule symmetrisch übereinanderliegende Trapeze, die das seitlich auslaufende Rücklicht zwischen sich nehmen. Zusammen mit den anderen Formendetails der Karosserie könnte eine Serienversion dem Dacia zu besserer Werthaltigkeit verhelfen. Längst nicht so aufwendig, schließlich ist hier schon die Serie erreicht, zeigt der Citroën DS 3 von 2009 sein orginellstes und damit schnell identifizierbares Designdetail: Die B-Säule gerät zur stehenden Flosse eines Delphins, die Glasfläche des hinteren Seitenfensters geht (fast) flächenbündig in die der Tür über, denn dazwischen legt sich der mattschwarze Türrahmen, um ja nicht aufzufallen, was den Effekt der Flosse heftig stören würde. Und die konstruktiv notwendige B-Säule ist ganz nach innen verlegt und schwarz gehalten hinter der Glasfläche. Diese Make-up-Methode nutzt der Concept-Car von BMW, der i3 von 2011, ausgiebig. Über den Hinterrädern ähnelt er konzeptionell dem Duster unter Verzicht auf die in die Seiten geführten Schlußlichter, nutzt dagegen fast die gesamte Seitenansicht für eine zusammenhängende Glasfläche, die überall dort, wo sie konstruktiven Kontakt mit der Säule oder dem Türrahmen hat, diese schwarz abdeckt und selbst tief dunkel eingefärbt ist. Das schafft die optisch zusammenfließende Wirkung des Seitenfensterglases.

Was noch von der Serie befreite Konzepte als Idee umsetzen können, wird vom Fließband in trivialerer Version zu den Kunden geschickt. Wenn dann, wie bei der kleinen Klasse, der Rotstift mit am Design zeichnet, müssen Gestaltungsideen großer Vorbilder viel schwarze Schminke auftragen. Als Beispiel dient der Hyundai i10 von 2013. Hier kommen viele Schwarzanwendungen

330. Der Citroën DS3 wird 2009 der Meister der Haifischflosse, man muß nur die B-Säule kappen. (Photo: Citroën.)

331. 2013 will der kleine Hyundai die Fenster größer machen, als sie sind – etwas peinlich. (Photo: Hyundai.)

332. Bei der BMW-Studie i3 kommt 2011 vieles zusammen: ein Hauch Fuego, ein Touch Duster, ein wenig Marzal. Geht nur, wenn Vorbilder vergessen sind. (Photo: BMW.)

333. 2014 verhelfen der Studie Biofore aus Finnland schwarze Farbe und abfallende Gürtellinie zur Aufmerksamkeit. (Photo: Biofore.)

334. Der Studie Honda Civic Tourer fliegt 2015 fast das Dach weg, so fließend weigert sich die C-Säule, die Gürtellinie anzurühren. (Photo: Honda.)

330. The 2009 Citroën DS3 masters the shark fin – you just need to cut back the B-pillar. (Photo: Citroën.)

331. In 2013 the little Hyundai tries to make the windows larger than they are – somewhat awkward. (Photo: Hyundai.)

332. The 2011 BMW study i3 combines many features: a touch of Fuego, a touch of Duster, a bit of Marzal. That only works if you've forgotten the models. (Photo: BMW.)

333. The black color and sloping beltline of the 2014 study Biofore from Finland draw attention to this model. (Photo: Biofore.)

334. The roof of the 2015 study Honda Civic Tourer almost flies away: The C-pillar never touches the beltline. (Photo: Honda.)

sized shutline between the undercarriage and the body. Not unlike a pleated bellows, it separates the bodywork, looking as if it could grow upwards. It is difficult to see a functional reason for it. Thus the black shutline is possibly misinterpreted, but in the long run it has given the Fuego an unmistakable look. The overemphasized separation of the body section can serve as a model when the designer tries out a similar separation between the roof pillars and the roof. When, thanks to a black shutline, the C-pillar is differentiated from the roof, the roof starts to float. The 2003 concept car Lancia Granturismo Stilnovo appeared very early, but was hardly better designed in subsequent years. No design after it achieved the clarity of the glass surface of the rear window, the metal panel of the roof and the C-pillar panel, all of them parts with a slight curvature. The 2008 BMW M1 Hommage had the smarts to combine the Fuego concept with that of the Lancia, because its flat silhouette was able to prolong this separation of the roof and undercarriage – which enhanced the lightness of the car body – instead of forming a continuous panel above the rear wheel if there had been no separation. In 1978 its predecessor merely had an elongated, black, scaled gill here for ventilating the mid-mounted engine, but paved the way for this development.

Of all the low-priced car brands, it was Dacia that presented its first, totally redesigned concept car in 2009, naming it Duster. Here the C-pillar is transformed into two halves, separated by a wide, reddish black shutline, located deeper in the direction of the wheel than the Lancia motif. Thus the halves of the C-pillar form symmetrically superimposed trapezoid shapes and, between them, the taillights, tapering off to the side. Together with other formal details of the bodywork, a production version might increase the Dacia's market value. By far less costly – after all, it's already being series-manufactured – the 2009 Citroën DS 3 sports its most original and thus easily identifiable design detail: The B-pillar turns into the upright flipper of a dolphin, the glass panel of the rear side window transitions (almost) flush-mounted into that of the door, for between them lies the matte black doorframe, careful not to stand out, which would definitely diminish the effect of the flipper. And the whole structurally necessary B-pillar has been moved inside and looms black behind the glass panel. BMW's concept car, the 2011 i3, makes extensive use of this makeup method. Above the rear wheels it resembles the Duster in conceptual terms, though it does not have the rear lights that extend into the sides; on the other hand, it uses almost the entire side profile for a continuous glass panel. Wherever this panel has structural contact with the pillar or the doorframe, it covers these in black and is itself tinted very dark. This creates the optical effect of merging with the side window glass.

Concepts that can be implemented come off the production line in a more trivial version and are passed on to customers. But if, as in the subcompact category, the design has been modified, the design features of the bigger models now need a lot of black makeup. An example is the 2013 Hyundai i10. Here black is used copiously. The rear windshield is meant to look substantial, so it extends into the sides with blind spots and masks part of the interior with black. A one-hundred-percent window thus loses almost 25 percent of its transparent glass surface. Almost as big as the i10, the 2014 study Biofore comes from Finland. All the glass side panels are skillfully flush-mounted. The roof does not come into contact with the C-pillar, a typical feature of the Lancia. The lines outlining the windows of the side elevation, however, are very clumsy. The 2015 Honda study Civic Tourer, thanks to a slightly declining station wagon roof, is able to virtually eliminate the C-pillar, for the glass of the third side window extends as far as the rear windshield. Black helps emphasize the difference between the roof and the glass, and the idea was retained even by the production model. The design of the 2015 Opel Astra worked along similar lines. Here

335. 2015 kann der Opel Astra gerade noch die Dachlinie und den C-Säulenstummel miteinander in Einklang bringen. (Photo: Opel.)

336. Etwas billig wirkt die schwarze Trennlinie am Fuß der C-Säule des Aston Martin DB 11 von 2017. (Photo: Aston Martin.)

337. 2017 kann der Suzuki Swift die C-Säule mit dem hinteren Türgriff kombinieren: ein Pseudo-Funktionsargument – fast glaubwürdig. (Photo: Suzuki.)

338. Der Opel Crossland X kombiniert 2017 Chromstreifen, schwarze Trennfläche und Hecklichtglas fast gekonnt, wären die Details nur nicht so plump. (Photo: Opel.)

339. 2016 macht der Lexus RX 450H alles wie immer bei Lexus: feinlinig, aber wo bleibt die Eleganz? (Photo: Lexus.)

340. Der einst kleine Nissan Micra bläst sich 2017 auf, schont weder Blech noch Lichterglas und macht mit bei der amputierten C-Säule. (Photo: Nissan.)

zum Tragen. Das Heckfenster soll großzügig wirken, greift mit Blindflecken in die Seiten und klebt mit Schwarz alles ab, was der Innenausstattung geschuldet ist. Aus hundertprozentigem Fenstermotiv werden so knapp 75 Prozent transparente Glasfläche. Fast gleich groß wie der i10 ist die Studie Biofore aus Finnland von 2014. Gekonnt sind alle Glasflächen der Seiten flächenbündig. Das Dach verzichtet auf Kontakt mit der C-Säule, was ein Lancia-Motiv ist. Nur sehr ungelenk sind die Fensterumrißlinien der Seitenansicht. Die Honda-Studie Civic Tourer von 2015 kann es sich dank leicht fallendem Kombidach leisten, die C-Säule so gut wie aufzulösen, denn das Glas des dritten Seitenfensters zieht sich bis ans Heckfenster in die Länge. In Schwarz wird nachgeholfen zur Betonung des Unterschieds von Dach und Glas und als Idee bis in die Serie durchgehalten. Der Opel Astra arbeitete 2015 gestalterisch ähnlich. Hier läuft die Gürtellinie an der C-Säule hoch, stoppt aber vor dem Dachkontakt. Die verchromte Dachlinie senkt sich nach hinten ab und betont so die Weiterführung der fallenden Dachkontur. Der Spalt zwischen dem Dach und der natürlich schwarzen C-Säule kämpft mit dem Heckfenster, weil die Oberkante der gestutzten C-Säule länger ausläuft als das Ende der Dachlinie, denn diese mußte Rücksicht nehmen auf die Höhe des Heckfensters. Im Design-Ansatz verstanden, fehlt in der Ausführung jene Perfektion des Lancia-Vorbilds.

Der BMW i3 als Konzeptauto bekam 2015 seinen Serienachfolger, konnte aber nur die aufgelöste C-Säule in den Alltag übernehmen. Anstelle der großen, abgedunkelten Glasfläche der Seitenansicht bleiben einzelne Fensterflächen übrig. Die verlangen vom Auge viel Toleranz, weil das hintere Seitenfenster an der B-Säule einen großen Knick nach unten macht, der noch in der Türfläche liegt und so eine durchgehende Gürtellinie heftig unterbricht, denn diese beginnt indirekt schon als schwarzer, dicker Streifen an den vorderen Kotflügeln und unterbricht die A-Säule an der Wurzel – ein Motiv, das vom Renault Fuego stammt und auch vom Toyota Aygo von 2016 in Schwarzgraphik nachgemalt wird. Das hintere Seitenfenster liegt tiefer, was an dem funktionalen Vorteil liegen könnte, daß der schräg rückwärts blickende Fahrer besser rangieren kann. Dieses Argument könnte einst Citroën-Designer veranlaßt haben, beim C2 1990 ähnliche Linienstörungen der Seitenfenster vorzunehmen. Eine Mischung aus BMW und Honda ist die verchromte Dachlinie des Lexus RX 450h von 2016. Chrom ist dann sehr hilfreich, wenn das Auto schwarz oder dunkel lackiert ist. Das unterstützt das Motiv des schwebenden Daches, auf das selbst dieser SUV nicht verzichten will und sich in die Reihe der Modeabhängigen begibt. In dieser Gruppe muß sich auch der Nissan Micra einordnen lassen. Seine Vorgänger waren noch zurückhaltend mit Motiven, die für einen Kleinwagen unpassende Kleider waren. Das Modell von 2017 wächst

the beltline runs up the C-pillar, but stops before it contacts the roof. The chrome-plated roofline moves downward toward the rear, thus emphasizing the continuation of the declining contour of the roof. The gap between the roof and the naturally black C-pillar struggles with the rear windshield, because the upper edge of the shortened C-pillar extends farther than the end of the roofline since the latter had to take into account the height of the windshield. While this was understood in the initial design, its final implementation lacks the perfection of the car's Lancia model.

In 2015 the concept car BMW i3 went into mass production, but was able to retain only the eliminated C-pillar in its everyday version. Instead of the large, darkened glazed area of the side elevation, only a few window areas still remain. These demand a great deal of visual tolerance, because the rear side window bends sharply downward at the B-pillar; this kink is still present in the door panel and thus abruptly interrupts a continuous beltline, for the latter already indirectly begins as a thick black stripe on the front fenders and interrupts the A-pillar at its root – a motif originally found in the Renault Fuego and also copied by the 2016 Toyota Aygo in black graphics. The rear side window is lower down, perhaps due to the functional advantage that this allows the driver who is looking backward diagonally to maneuver more easily. It is this argument that might once have caused Citroën designers to put in similar breaks in the lines of the side windows of the 1990 C2. The chrome-plated roofline of the 2016 Lexus RX 450h is a mixture of BMW and Honda. Chrome is very helpful when the car is black or painted a dark color. This supports the motif of the floating roof, which even this SUV refuses to forego, thus joining the ranks of the fashion-conscious. The Nissan Micra is part of this group. Its predecessors were still reticent in their

335. In the 2015 Opel Astra the roofline and the stump of the C-pillar are barely brought into line. (Photo: Opel.)

336. The black separation line at the foot of the C-pillar of the 2017 Aston Martin DB 11 looks a bit cheap. (Photo: Aston Martin.)

337. In 2017 the Suzuki Swift is able to combine the C-pillar with the rear door handle: a pseudo-function argument – almost credible. (Photo: Suzuki.)

338. The way the 2017 Opel Crossland X combines chrome strips, black separating surface and taillight glass is almost masterly, if only the details were not so clumsy. (Photo: Opel.)

339. In 2016 the Lexus RX 450H has the same old Lexus features: The lines are delicate, but what happened to elegance? (Photo: Lexus.)

340. In 2017 the formerly small Nissan Micra becomes inflated, spares neither metal nor headlight glass and joins the trend of the pinched C-pillar. (Photo: Nissan.)

und strebt nach allen greifbaren Modetendenzen, wie die überzogene Flankengestaltung oder die ausgefransten Hecklichter. Und selbstverständlich muß das Dach schweben. Dann erinnert man sich an einen Alfa 156 mit verstecktem Türgriff an der Hintertür und spendiert dem Micra diesen für die dicke Fuge zwischen Dach und C-Säule. Beim Alfa 147 und 156 machte das einen coupéähnlichen Eindruck, beim Micra wurde es zum Gag.

Das gleiche Rezept übernimmt auch der Suzuki Swift in seinem Modell von 2017, ohne die Eleganz eines Alfa 156 zu erreichen. Dort fällt der Griff kaum auf, hier liegt der in der schwarz aufgelösten C-Säule im Türbereich und zeichnet im Bereich der C-Säule die Griffmulde nach, was kein Versteckspiel ist, sondern Aufmerksamkeit ernten will. Warum überhaupt? Der Swift ist doch ansonsten ein unaufgeregter Entwurf. Die geschwärzten A-Säule und B-Säule, schon seit Jahrzehnten übliche Lackiervariante, tragen das Dach, betonen die Linie der Regenrinne und schaffen so Ähnlichkeiten zu Konkurrenzmodellen wie einem Skoda Fabia von 2007 oder bei den Größeren wie dem Saab 9-5 von 2011. Aber das sind unbedeutende Parallelen und weniger auffällig als Türgriffe an geschwärzten C-Säulen. Der Opel Crossland von 2017 ist einerseits ein Astra-Adept, löst diesmal das Dach an der Gürtellinie von der C-Säule und betont das Motiv mit kräftiger Chromlinie. Wie bei einem Dacia Duster wird die Fuge nicht geschwärzt, sie nimmt das dunkelrote Glas des Rücklichts auf. Und so wird aus einer guten Idee eine um sich greifende Mode. Gut, wenn gut gemacht.

7.11. Die Form braucht Farbe, die Technik nicht

»Silberpfeile«, dieser Begriff beschreibt Farbe, läßt sofort an Mercedes-Rennwagen denken – und ist kein aufgetragener Lack, sondern Materialton. In den 1930er Jahren rollten Stuttgarter Renner auf die Waage zur Abnahme vor dem Rennen. Ergebnis: einige Kilo zu schwer. Mercedes-Rennleiter Fritz Nallinger gab kurz Order: runter mit dem Lack. Ergebnis: Gewicht gehalten, die Aluminiumkarosserie trat hervor – sie glänzte silbern. Beim Thema Zweifarbigkeit kann auch die »Farbe« des Materials gelten, wenn zum Beispiel beim Mercury Marmon Herrington Modell 89M von 1948, einem Vertreter der »Woodies«, Holz und Lack sich zweifarbig geben. Im Deutschland der Nachkriegszeit war der bekannteste »Woody« der DKW Kombi F89 S von 1951. Zweifarbige Automobile gibt es fast so lange, wie es Autos gibt. Der Amilcar von 1922 hatte den Materialmix von Holz und Aluminium, der Auburn 8-88 von 1928 eine Flächenaufteilung in zwei Lacktönungen, die zum Markenzeichen hätte geraten können. Das blieb dem Bugatti Royale von 1931 und einigen anderen Modellen der Marke vorbehalten. In starkem Kontrast von Schwarz zu Gelb, Rot oder Blau legte sich die schwarze Lackierung von der seitlichen Oberkante der Motorhaube in kühnem Schwung in die Flanken, was beim Royale mit 6,5 m Länge besonders großzügig wirkte. Bugatti ist das Beispiel für frei geformte Farbflächen, die keine Rücksicht auf Karosserieelemente nehmen, wenn Dach und Unterbau oder Kotflügel und restliche Karosserie unterschiedlich lackiert sind. 1949 zeigte das Käfer-Cabrio die disziplinierte Anwendung zweier Farben: Schwarz für das gesamte Auto, passend zum schwarzen Stoffverdeck, Rot die klar begrenzte Fläche der Seiten unterhalb der Gürtellinie.

Eine erweiterte Freiheit der Farbflächenaufteilung wäre das Dach, das mit Teilen des Unterbaus korrespondiert. Der Buick Special von 1955 nutzte die verchromte »schnelle Linie« als Trennlinie zweier Farben, um eine davon im Dach wieder aufleben zu lassen, ein Standardrezept bei sehr

341. 1955 wirkt es schon. Wer Weiß als zweite Farbe wählt, holt Frische auf die Karosse: wie beim Buick Special. (Photo Wikimedia.)
342. Holz kann 1948 auch zur zweiten Farbe werden, wie beim Ford Mercury Herrington Model 89M. Solche Autos sind dann die »Woodies«. (Photo: Crittenden.)
343. 1955 schon üblich: Flankenprägung farbig absetzen, wie bei dem Gaylord Gladiator. (Photo: Imgur.)
344. Zwei Farben können 1949 Ruhe schaffen, wenn sie bauteilweise angewendet werden – und vielleicht bei den Felgen wieder auftauchen, siehe VW Käfer Cabrio. (Photo: Volkswagen AG.)

use of motifs that were inappropriate for a subcompact car. The 2017 model strives to adopt each and every fashionable trend, such as the exaggerated design of the flanks or scraggy-looking taillights. And it goes without saying that there has to be a floating roof. Then the designer remembers an Alfa 156 with a concealed rear door handle and slaps one on the Micra for the hefty gap between the roof and the C-pillar. This made the Alfa 147 and 156 look like coupés, but in the Micra the feature turned into a gag.

The Suzuki Swift follows the same recipe in its 2017 model, without achieving the elegance of an Alfa 156. There, the handle is barely conspicuous, while here it is located in the black C-pillar in the door area and in the area of the C-pillar traces the outline of the handle recess, a bid for attention. Why is this necessary? After all, the Swift is an otherwise bland design. The blackened A-pillar and B-pillar – in the kind of color that had been standard for decades – support the roof, emphasize the line of the rain groove and thus create similarities with competing models like a 2007 Skoda Fabia or larger cars like the 2011 Saab 9-5. But those are insignificant parallels, and less conspicuous than door handles on blackened C-pillars. The 2017 Opel Crossland is, on the one hand, similar to the Astra, but does not detach the roof from the C-pillar at the beltline, and emphasizes the motif with a forceful chrome line. As in a Dacia Duster, the gap is not black, but picks up the color of the dark red glass of the taillight. And thus a good idea becomes a large-scale fashion trend. It's good if done well.

7.11. Form needs color, technology does not

»Silver arrows«. The expression describes color, and immediately makes you think of Mercedes race cars – and is not paint applied to the exterior, but the shade of the material itself. In the 1930s Stuttgart racing cars rolled onto the scales to be weighed before the race. The result: a couple of kilos too heavy. Mercedes' racing manager Fritz Nallinger curtly ordered: Get rid of the paint. Result: the weight was kept down, the aluminum bodywork emerged – with a silver glow. When it comes to two-tone cars, the »color« of the material may also count, for instance, when in the 1948 Mercury Marmon Herrington 89M, one of the »Woodies«, wood and paint have different colors. In postwar Germany the best known »Woodie« was the 1951 DKW F89 S station wagon. Two-tone cars have existed almost as long as there have been cars. The 1922 Amilcar mixed wood and aluminum, and in the 1928 Auburn 8-88 the exterior surface was divided into two shades of color that might have become its trademark. That was reserved for the 1931 Bugatti Royale and a few other models of the marque. In the sharp contrast of black with red, yellow or blue, the black paintwork swept boldly from the top of the side of the hood into the flanks, which looked particularly grand on the 6.5-meter-long Royale. Bugatti exemplifies free-form areas of color that do not take into account the elements of the bodywork when the roof and undercarriage or fenders and the rest of the body are painted in different colors. In 1949 the Beetle convertible showed the disciplined use of two tones: black for the entire car, matching the black fabric top, red for the clearly defined area of the sides below the beltline.

A roof that corresponds to sections of the underbody represents even more freedom in the distribution of areas of color. The 1955 Buick Special used the chrome-plated »fast line« as the

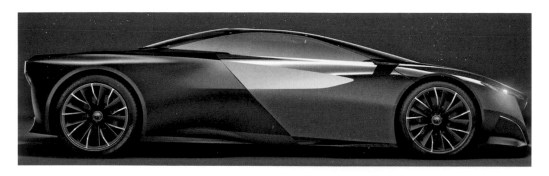

vielen Modellen der 1950er Jahre. Beim Buick paßte sich die Farbverteilung teilweise den Karosserieelementen an, so bei dem hinteren Radausschnitt. Freiere Anwendung der Zweifarbigkeit hatte der De Soto Firedome – eine Chrysler-Marke – von 1955. Die dynamisierte Umrißlinie war frei von baulichen Vorgaben. Der Gaylord Gladiator vom Designer Brooks Stevens war ein Concept-Car, dessen Flanke eine sogenannte »Cut-out-Fläche« an den Flanken hatte. Dieser Begriff benennt Flächenbereiche, die aus einer übergeordneten Fläche negativ, leicht konkav herausgeformt werden. Das kann zur Wiedererkennung eines Modells einer Marke betont werden. In England sind Zweifarblackierungen genau so beliebt wie weltweit. Hier bevorzugt die englische Zurückhaltung normalerweise dezente Kontraste, meist in der Mischung von schwarz und silbern. Im Detail wird manchmal nachgeholfen mit einer »Brushlinie«, einer handwerklich perfekt aufgetragenen Pinsellinie. Das gezeigte Beispiel des Armstrong Siddeley von 1956 war mutiger und nahm kontrastreich die Farbe Weiß, was in diesem Fall einem Hochzeitsauto nicht schadet. Interessant ist die Farbverteilung, weil sie sich nicht exakt an Karosserieelementen orientiert, etwa am vorderen Kotflügel. Was die Großen können, wollen die Kleinen auch haben. So konnte ein kleiner deutscher Transporter, der Tempo Wiking von 1955, die Zweifarbigkeit bescheiden anwenden, unten am Kühlerschnäuzchen, den Kotflügeln und dem Schweller, oben klassisch am Dach.

Vom Buick könnte der Austin A35 von 1957 seine Zweifarbigkeit gelernt haben. Er übernahm die bekannte »schnelle Linie« als Grenze von Hell und Dunkel, die sich an die blechgeprägte Vorgabe hielt. In Verbindung mit Weißwandreifen, einer bescheidenen Variante von Zweifarbigkeit bei unifarbenen Modellen, konnte sich der relativ kleine Wagen Respekt verschaffen. Ein deutsches Pendant wäre dazu der Glas Isar T 700 oder eine Lloyd Arabella. Der kleinere Lloyd Alexander war Basis für den Michelotti-Entwurf von 1968. Dieses Konzeptauto nutzte, wie viele speziell italienische Karossiers, die Zweifarblackierung. Anstelle großflächiger, von der Karosserieform abgeleiteter Farbaufteilungen reichten ihnen bandförmige Kontrastflächen mit kleinen Abweichungen, wie hier über dem Hinterrad, manchmal chromgefaßt. Aus der Hand von Michelotti stammt auch das formale Konzept des Triumph Herald. Die Zweifarbigkeit legt sich nicht auf Vorgaben der Flankenausbildung fest. Hier wird die ansteigende Linie der Farbtrennung genutzt, um eine gewisse Dynamik zu schaffen, was bei der unspektakulären Seitenansicht nur bedingt gelingt, sich aber an die disziplinierte Grundhaltung des Designs anpaßt. Angepaßt an die Grundhaltung des Designs hat sich auch der Citroën 2CV von 1983 in unüberbietbarer Weise. Einerseits kostete er den Dachschwung voll aus, andererseits zitierte er damit die Zweifarbigkeit einer Automarke am anderen Ende der Preisskala: dem Bugatti. Und die »Ente« kannte bei der Charleston-Version noch eine weitere Steigerung der Veredelung. Die rot ausgelegte Fläche an den Türen gab es auch mit achteckiger Geflechtstruktur in mildem Beige auf dunklem Grund.

Wie edel ein Materialmix die Zweifarbigkeit fördern kann, demonstrierte 2007 der Bertone-Entwurf des Fiat Barchetta. Diese Studie hatte großzügig verchromte Flächen, die im Kontrast zu den schwarzen Flanken stehen. In den schwarzen Bereichen schimmert die Kohlefaserstruktur durch

separation line between two colors, with one of them reappearing in the roof, a standard recipe in many 1950s models. In the Buick the color distribution was partly adapted to the elements of the bodywork, for instance in the case of the wheel cutout. The 1955 De Soto Firedome – a Chrysler marque – used the two-tone style more freely. The dynamized contour line was free of structural specifications. The Gaylord Gladiator by designer Brooks Stevens was a concept car whose flank had a so-called »cut-out area« in the flanks. This term refers to negative and slightly concave indentations in the bodywork impressed in a higher surface. The area may be accentuated for easy recognition of the model of a particular brand. In the UK, two-tone cars are just as popular as they are elsewhere in the world. Here British reserve normally prefers discreet contrasts, usually in a combination of black and silver. Sometimes a »brushline« is added, a line perfectly applied by hand with a brush. The example shown here, the 1956 Armstrong Siddeley, was bolder and chose white as a contrasting color, which in this case doesn't hurt, since this is a wedding car. The color distribution is interesting: It is not exactly in alignment with the elements of the bodywork, for instance on the front fender. The little cars want what the big ones have. Thus a small German van, the 1955 Tempo Wiking, opted for a modest two-tone color scheme – down below on the radiator, the fenders and the sill, above on the roof, a classic feature.

The 1957 Austin A35 might have learned its two-tone color scheme from the Buick. It adopted the well-known »fast line« as a border between light and dark, which was stamped into the metal. Along with whitewall tires, a modest variant of a two-color scheme in one-color models, this feature gained the relatively small car respect. A German counterpart would be the Glas Isar T 700 or a Lloyd Arabella. The 1968 Michelotti design was based on the smaller Lloyd Alexander. Like many specifically Italian car body manufacturers, the designer of this concept car used the two-tone color scheme. Instead of contrasting colors covering large areas, derived from traditional car body forms, the Italians were content with having ribbon-shaped contrasting areas with small deviations, like the one pictured here above the rear wheel, sometimes framed in chrome. Michelotti also created the formal concept of the Triumph Herald. Here the ascending line of color separation is used to create a certain dynamic, which is only partly successful due to the unspectacular side view, but is in line with the disciplined tenor of the design. In an unparalleled way, the 1983 Citroën 2CV is also in keeping with the basic tenor of the design. On the one hand it has made full use of the sweep of the roof, while on the other it thus quotes the two-tone color scheme of a marque at the other end of the price range: the Bugatti. And the »Duck« in its Charleston version experienced a further refinement: The area on the doors that was lined with red also came with an octagonal mesh in a mild beige on a dark background.

In 2007 the Bertone design of the Fiat Barchetta demonstrated how a mix of materials can ennoble the appearance of a two-tone car. This study model had generously chrome-plated surfaces that contrast with the black flanks. In the black areas the carbon fiber structure shimmers through and at the same time enhances the appearance of the material mix. Black and chrome are continued down to the design of the rims and emphasize the consistency of the overall design. It is regrettable that due to the cost factor bodywork whose surface is largely chrome cannot be mass-produced. For as a surface finish, chrome lasts longer than paint. You can see that in junkyards, where rust bleeds through dull paintwork, while chrome still flashes bright. But it need not always be chrome – it could be copper, too. Why the change? Perhaps because there's been a gradual change from the combustion engine to the electric drive. And of course an electric drive has copper wire coils. This is how copper probably wormed its way into filigree rim de-

345. The 2012 Peugeot study Onyx has the bright idea of painting the front section copper. After decades of chrome this is an attractive feature that can't help attracting attention. (Photo: Peugeot.)
346. The 2017 Bugatti Chiron draws on its own brand's classical tradition of two-tone paintwork that ends in a curve. This makes for ongoing brand recognition. (Photo: Bugatti.)
347. In 1955 a hard-working Cinderella becomes a suburban beauty after putting on white makeup as a Tempo Viking. (Photo: Rezbach.)
348. The cheapest car, in 1983, acts like the top of the range: The Citroën 2CV – the Duck – tries to rival the Bugatti. That's no crime, because it's funny. (Photo: Citroën.)
349. The color distribution of the 1956 Armstrong Siddeley continues to be elegant – after all, we're off to a wedding. (Photo: CWeddingcars.)
350. The blue band extends the short wheelbase of Michelotti's 1958 Lloyd study. (Photo: Michelotti.)
351. In 2007 Bertone shows that thanks to chrome a two-tone design is feasible and practical for the Fiat Barchetta, for chrome is more durable than paint. (Photo: Bertone.)

und veredelt zusätzlich den Material-Mix. Schwarz und Chrom werden weitergeführt bis in das Felgendesign und betonen die Stimmigkeit des gesamten Entwurfs. Bedauerlich ist, daß eine Karosserie mit großflächigem Chromanteil aus Kostengründen kein Vorschlag für eine Serie sein kann. Denn bezogen auf Oberflächenvergütung ist Chrom ausdauernder als eine Lackierung. Das studiert man auf Autofriedhöfen, wo matter, rostiger Lack den automobilen Tod zeigt, Chrom aber noch aufblitzt. Aber es muß nicht immer Chrom sein, es darf auch Kupfer werden. Zu rätseln ist, woher dieser Wechsel kommen könnte. Eine gar nicht abwegige Vermutung mag der schleichende Wechsel vom Verbrennungsmotor zum Elektroantrieb sein. Und der hat nun mal Kupferdrahtwicklungen. So könnte Kupfer sich in filigranes Felgendesign oder feingliedrige Grills eingeschlichen haben. Oder aber eben in die vordere Hälfte der Peugeot-Studie Onyx von 2012, deren Farbwechsel in nach hinten geneigter Linie erfolgt, was dynamisch wirkt und Vorgabe für weitere Peugeot-Modelle auch der Serien ist, wenn es sich um PS-gestärkte Versionen handelt. PS-gestärkt bis zum Abwinken, bietet der Bugatti Chiron von 2017 die klassische Trennlinie zweier Farben wie in den 1930er Jahren, hier in gestalterischer Gemeinsamkeit mit der Charleston-Ente. Welch ein Kontrast!

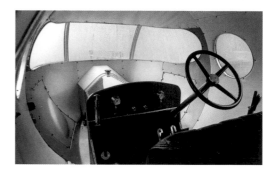

7.12. Form und Technik müssen sich einig sein

Kraft, umgesetzt in kinetische Energie, muß kontrolliert werden. So sind die Geschwindigkeit zu messen, der Benzinstand – oder bald die Batteriekapazität –, der Ölstand, die Kühlwassertemperatur, vielleicht noch die Uhrzeit und die Außentemperatur, eine gezogene Handbremse, eingeschaltetes Licht, die Klimaanlage, die Blinker. Ist das Auto brandneu und die Aufpreisliste voll ausgeschöpft, können weitere Anzeigen folgen: ESP, Totwinkelwarner, Verkehrszeichenerkennung Head-up-Displayanzeige. Unberücksichtigt bleiben dabei: Radio, Navigationsgerät und Smartphoneinsatz, dieser nur im Freisprechmodus. Wer die Summe aller Elemente vor dem Fahrer aufbaut und alles noch Armaturenbrett nennt, macht das nicht zur Beschreibung eines Gegenstands mit Materialhinweis, sondern nutzt den alten Begriff, so wie der Kotflügel kein Flügel mehr ist und höchst selten vor Kot schützt, seit das Pferd kein Verkehrsteilnehmer mehr ist. Trotz allem: Das Armaturenbrett war am Anfang ein Brett, auf das wenige Anzeigen oder Uhren aufgeschraubt waren. Da sie Uhren waren und diese Zifferblätter haben, die seit Jahrhunderten fast nur schwarze Ziffern auf weißem Grund kennen, so wie Zeitungen zur optimalen Lesbarkeit gedruckt waren und werden, wurden sie auch entsprechend im Auto verwendet. Selbst eine üppige Gestaltung, soweit diese angemessen war, hatte sich zurückgehalten bei der typographischen Auslegung von Ziffern und Buchstaben. Hier unterscheidet sich die Uhrensammlung im Auto von jenen moderner Armbanduhren. Die Ablesbarkeit ist kaum gefragt, der Variantenreichtum dagegen bis hin zu peinlicher Auslegung dessen, was Design ist. Beim Automobil sind schnell erfaßbare Anzeigen im speziellen Fall eine lebensnotwendige Eigenschaft. Die Entwicklung des Armaturenbretts belegt

352. 1929 keimt der Wille, Armaturen auch zu formen, selbst wenn sie aus Holz sind. Nicht umsonst spricht man ja vom Armaturenbrett, wie hier beim Isotta Fraschini Tipo 8a. (Photo: Phantom.)
353. Fast wie im Klub: Gestepptes Leder ist gerade bei offenen Autos besser als Stoff. Hier ein Woods Electric-Style 214a Queen Victoria Brougham von 1905. (Photo: Wallpaperup.)
354. 1914 hatte der Alfa 40-60HP Ricotti von Castagna im Gegensatz zur revolutionären Form nichts Passendes für den Innenraum. (Photo: Shorey.)
355. Nichts ging 1949 über Gemütlichkeit für deutsche Autofahrerseelen beim Kauf des VW Export. (Photo: Volkswagen AG.)
356. 1955 hatte die Cadillac-Studie Eldorado Brougham wegen des Bildschirms fast schon heutiges Niveau. (Photo: GM.)
357. Der Adler Trumph Junior kann 1935 als eines der ersten Beispiele des »Wrap-around«-Armaturenbretts gelten. (Photo: Vintagedrivingmachine.)

352. In 1929 there is a beginning trend to design the instruments, even if they are made of wood. Pictured here is the instrument panel of the Isotta Fraschini Tipo 8a. (Photo: Phantom.)

353. Almost like being at the club: Particularly in open cars, quilted leather is better than fabric. Pictured is a 1905 Woods Electric-Style 214a Queen Victoria Brougham. (Photo: Wallpaperup.)

354. In 1914, in spite of its revolutionary form, the Alfa 40-60HP Ricotti by Castagna had nothing suitable in mind for the interior. (Photo: Shorey.)

355. Cozy comfort – the top selling point for German car buyers looking to purchase the 1949 VW Export. (Photo: Volkswagen AG.)

356. Because of the screen, the 1955 Cadillac study Eldorado Brougham was almost up to today's standards. (Photo: GM.)

357. The 1935 Adler Trumpf Junior may be considered one of the first examples of the »wraparound« display panel. (Photo: Vintagedrivingmachine.)

sign or graceful grilles. Or else into the front half of the 2012 Peugeot study Onyx, whose color changes in a sweep toward the rear, which has a dynamic effect and is a harbinger of things to come for future Peugeot models, including powerful series-produced versions. The 2017 Bugatti Chiron, powerful as all get-out, has the classic 1930s separation line between two colors, a design feature it shares with the Charleston Duck. What a contrast!

7.12. Form and technology must be in agreement

Power, transformed into kinetic energy, must be monitored. We need to check the speed, the fuel level – or, soon, the battery capacity –, the oil level, the coolant temperature, perhaps even the time of day and the ambient temperature, whether the handbrake has been applied, whether the headlights are on, the air-conditioning, the turn signals. If the car is brand-new and comes with a complete set of options, we may need to check other displays as well: ESP, blind-spot alert, traffic sign recognition, head-up display. We haven't even mentioned the radio, the travel-assist system and smartphone startup, the latter only in hands-free mode. We call the sum of all these elements in front of a driver a dashboard. Just as a mudguard now very rarely protects us from mud, the term dashboard no longer refers to the actual board it initially was, which had a few gauges or dials screwed onto it. Since these were clocks and the dials of clocks have for centuries almost always had black numbers on a white background, for easier readability like newspapers, these were also used in cars. Even lavish designs had been conservative when it came to styling numbers and letters. This is where the collection of dials in a car differs from modern wristwatches. Readability is hardly a priority, while on the other hand there is a plethora of variants, and scrupulous attention to design. In a car, easily identifiable displays may in some cases be a matter of life and death. The evolution of the dashboard has demonstrated this time and time again, but at the beginning of the car age it mattered far less than the appearance of the car's interior. At one time the standard of comparison was the boudoir rather than the kind of sober look one might expect in an automobile. Inside the 1905 Woods Electric Style 214a Queen Victoria Brougham – what a name for a model! – you felt like you were in a salon: diamond quilting everywhere, roller blinds on the windows, a flower vase, with only an inconspicuous lever somewhere, presumably the emergency brake. On the other hand, the Alfa Romeo 40-60 HP of Count Marco Ricotti, built by Castagna in 1913, was a sober example of a true dashboard, having two small dials and a pump on it, presumably for controlled lubrication.

How quickly, a mere 16 years later, the dashboard had expanded! It now had a padded upper edge and a large number of levers and indicators. As early as 1929, the Isotta-Fraschini Tipo 8a even had a steering wheel with several gas and ignition settings. And even at the time the driver's feet had to control three pedals plus a button for the high beam. Another striking feature were the graceful ventilation louvers for warm air during winter drives. The driver's compartment of a 1935

dies mit jedem Modell neu, war aber zu Beginn der Mobilität nebensächlicher als die Frage, wie es im Auto aussieht. Da war das Boudoir eher der Maßstab als automobilgeprägte Nüchternheit. So sah es im Woods Electric Style 214a Queen Victoria Brougham – welch Modellname! – von 1905 aus wie in einem Salon: Rautensteppung allüberall, Fensterrollos, Blumenvase, nur unauffällig irgendwo ein Bedienungshebel, vermutlich die Handbremse. Dagegen war der Alfa Romeo 40-60 HP des Grafen Marco Ricotti, gebaut 1913 von Castagna, ein nüchternes Beispiel für das wahre Armaturenbrett, bestückt mit zwei kleinen Uhren und einer Pumpe, vermutlich zur kontrollierten Schmierung.

Wie schnell hatte sich nach nur 16 Jahren das Brett gestreckt, gab es eine gepolsterte Oberkante und eine Vielzahl von Hebeln und Anzeigen. Selbst das Lenkrad mit mehreren Stellfunktionen für Gas- und Zündeinstellung kannte bereits 1929 der Isotta-Fraschini Tipo 8a. Und die Fußarbeit mußte schon damals drei Pedale beherrschen nebst Knopf für das Fernlicht. Auffallend waren auch die zierlichen Lüftungsgitter für Warmluft bei winterlichen Ausfahrten. Ausreichend, aber karg in der Ansicht zeigte sich der Fahrerplatz des Adler Junior von 1935. Wie der Isotta trug er eine gepolsterte Oberkante, die ohne Unterbrechung in die Oberkante der Tür gezogen war, einer der ersten Fälle des sogenannten »Wrap-around«-Armaturenbretts. Deshalb englisch bezeichnet, weil besonders seit der Nachkriegszeit die Amischlitten diese Designlösung bevorzugten, um die Fahrerumgebung großzügig als geschlossene Einheit zu gestalten. Ganz anders war der erste Nachkriegs-Käfer von 1949 in der Exportversion. Strapazierfähige Stoffe, ein Armaturenbrett der perfekten Symmetrie für Links- oder Rechtssteuerung ohne bauliche Veränderung des blechgeprägten Armaturenbretts wie bei der Standardversion. Der Luxus aber war die Sofarolle, sinnvollerweise nicht fixiert, sondern als loses Zubehör, bereit für jede frei wählbare bequeme Lage im Fond. Luxus ist steigerungsfähig, besonders in den USA. Hier pflanzte man dem Concept-Car Cadillac Eldorado Brougham von 1955 TV-Gerät und Radio in die Rückseite der Frontlehnen, Platz genug gab es ja. Flachbildschirme an Kopfstützen für kleine Passagiere sind daher nichts Neues. Wo beim Caddy gepolsterte Flächen üppig Verwendung fanden, da blieb 1956 ein Mercedes W 180 und 128, auch bekannt als 220 S, dem Massivholz treu und baute aus einem Stück einen kompletten Armaturenträger, denn »Brett« wäre wohl der falsche Begriff gewesen. Die »Wrap-around«-Methode zeigte hier den perfekten Übergang. Wie aufwendig mag die Arbeit gewesen sein, Holz in dieser unterschiedlichen Ausformung für die Serie aufzubereiten!

Wie man sich Arbeit sparen und handwerkliche Leistung durch die Üppigkeit von Schalter und Uhren kompensieren kann, demonstrierte der Concept-Car Chrysler Diablo von 1957 und wertete den Fahrer so auf, daß er sich wie ein Pilot fühlen durfte. Hier zeigte die Panoramscheibe, daß ein »Wrap-around«-Design sinnvoll und ästhetisch vollkommen wirkt, wenn es konsequent ausgeführt wird. Beim Diablo stolperte die Polsterung in die Türverkleidung. Das passierte leider auch dem Citroën ID/DS, hier in der Variante von 1959. Der Designer Flaminio Bertoni, von Haus aus Bildhauer, hatte für die Armaturen nicht das perfekte Händchen. Da wurden aus runden Anzeigen

358. 2015 macht der Bentley EXP10 Speed-6 alles teuer, aber erinnert stark an eine Corvette der 1950er Jahre mit der ansteigenden Mittelkonsole. (Photo: Bentley.)

359. Handarbeit war 1956 gefragt, als sich Massivholz wie gepolstertes Leder gab, wie beim Mercedes-Coupé W180-128. (Photo: Daimler.)

360. 1957 sind viel Show und flirrende Oberfläche gefragt, so wie beim Chrysler Diablo. (Photo: Chrysler.)

361. Der Citroën DS von 1959 hat bereits das zweite Armaturenbrett, das konventioneller wurde, aber das Einspeichenlenkrad blieb – zum Glück. (Photo: Citroën.)

362. 2016 dürfen auch Fondgäste wie der Beifahrer vornehm chauffiert werden: teuer, edel, üppig – einem Bentley Mulsanne Speed angemessen. (Photo: Bentley.)

358. The 2015 Bentley EXP10 Speed-6 is a luxury car in every way, but strongly reminds us of a 1950s Corvette because of its rising center console. (Photo: Bentley.)

359. Craftsmanship was in demand in 1956, when solid wood presented itself as padded leather, as in the Mercedes Coupé W180-128. (Photo: Daimler.)

360. In 1957 lots of show and shimmering surface are in demand. Pictured here is the Chrysler Diablo. (Photo: Chrysler.)

361. The 1959 Citroën DS already has the second display panel, which has become more conventional, but – luckily – still retains its single-spoke steering wheel. (Photo: Citroën.)

362. In 2016 back-seat guests – like the front-seat passenger – get first class treatment: expensive, classy, lavish – commensurate with a Bentley Mulsanne Speed. (Photo: Bentley.)

Adler Junior was adequate but looked spartan. Like the Isotta, its dashboard had a padded upper edge that connected seamlessly with the upper edge of the door, one of the first instances of a so-called »wrap-around« dashboard. It was called an English dashboard because especially since the postwar period, the big Yankee cars had favored this design solution, where the driver's environment was designed as a spacious closed unit. Not so the first Beetle in the 1949 export version. Durable fabrics, a perfectly symmetrical dashboard for left- or righthand steering without structural modification of the stamped steel dashboard as in the standard version. But its luxury feature was the bolster, which sensibly was not attached, but was a loose accessory and could be used for any convenient location in the back of the car. Luxury is capable of improvement, especially in the U.S. There, they installed a TV set and radio in back of the front seats of the 1955 concept car Cadillac Eldorado Brougham – there was certainly enough room. That is why flat-screen TVs on the headrests for little passengers are nothing new. Whereas in the Caddy upholstery was used in abundance, the 1956 Mercedes W 180 and 128, a.k.a. the 220 S, stayed with solid wood and built a complete instrument panel – for »board« would probably have been the wrong term – out of a single piece. The »wrap-around« method here showed a perfect transition. How labor intensive this special process of shaping the wood for series production must have been!

It is possible to save hours of labor and making up for them by using lavish switches and dials, a fact demonstrated by the 1957 concept car Chrysler Diablo, where the driver's role was enhanced to the point where he felt like a pilot. A panoramic windscreen showed that a »wrap-around« design makes sense and looks aesthetically complete if it is implemented logically. In the Diablo the upholstered padding of the door didn't pan out. The same was unfortunately also true of the Citroën ID/DS, pictured here in the 1959 variant. The designer, Flaminio Bertoni, originally a sculptor, lacked the expertise for designing dashboards. Round displays turned into rectangles or, as in the case of the speedometer, truncated sections of circles. Only the steering wheel was ingeniously simple, because the spoke that curved into the steering column proclaimed its safety even visually, though it was not suitable for the airbags that came later. How restrained but appealing, on the other hand, was the study BMW Z18 built in 2000. Because they were not affected by fashion, three dial-type gauges, black on white displays, and a padded »wrap-around« design were good for the duration. Along the same lines was the »wrap-around« design of the 2011 Audi A6; it was restricted to the top part of the car, leaving lots of room for the lavish design of all parts of the car while adding a great deal of upholstery, while wood was used only sparingly. A style without wood but small amounts of matte aluminum and a lot of free space is feasible when the outsider Tesla designs the driver's compartment. Its 2015 Model X probably has the largest display screen of its period, which makes sense when practically all switch functions can be made to be touch-sensitive instead of the usual click-clack.

rechteckige oder, wie beim Tacho, angeschnittene Kreispartien. Lediglich das Lenkrad war genial einfach, weil die in die Säule gekrümmte Speiche schon optisch Sicherheit deklarierte, aber für spätere Airbags nicht geeignet war. Wie zurückhaltend, aber ansprechend war dagegen die Studie BMW Z18 von 2000. Drei Rundinstrumente, Anzeigen schwarz auf weiß, gepolstertes »Wrap-around«-Design, waren, weil modisch unberührt, gut für eine kleine Ewigkeit. In diese Richtung tendierte auch das »Wrap-around«-Design vom Audi A6 von 2011; es zieht sich auf die oberste Etage zurück und läßt der Üppigkeit aller Partien viel Raum, zur Sicherheit aber viel Gepolstertes und Holz nur als zurückhaltende Applikation. Kein Holz, aber wenig mattes Aluminium und viel freie Fläche sind dann machbar, wenn der Außenseiter Tesla den Fahrerplatz gestaltet. Das Model X von 2015 dürfte den größten Bildschirm seiner Zeit haben, sinnvoll, wenn möglichst alle Schalterfunktionen berührungsempfindlich ausgelegt werden können anstelle der üblichen Klick-Klacks.

England, ein Meister der Klubatmosphäre, läßt diese Einrichtungsidee bei seinen Luxusmarken aufleben, auch wenn VW bei Bentley schon längst die Hand darauf gelegt hat. Der EXP, eine Studie von 2015, schwelgt in Rautensteppung, verwirrt das Auge mit Glaskristall und bietet Fahrer wie Beifahrer ein eigenes »Wrap-around«-Gefühl. Sein Serienlimousinenableger kann es auch üppig, hier wieder mit Rautensteppung. Und das Champagnerglas wird aus der Mitte des Fauteuils gereicht – aber nur für Fondgäste des Bentley Mulsanne Speed von 2016. Wie anders das Ambiente, wenn Designer bei Honda den zukünftigen Fahrerplatz gestalten. Die Studie NeuV von 2016 nutzt neueste Technik der Bildschirmanzeige, die sich grenzenlos gibt und alles in sich vereint, was Kontrolle und Bedienung eines Automobils notwendig macht. Dazu die Denksportaufgabe, wie ein Lenkrad anders als üblich gestaltet werden kann: einfach einen Ring durch zwei Führungsklammern laufen lassen. Zurück in der Welt von heute bietet der Opel Ampera von 2016 jene Armaturenauslegung der bekannten Art: »Wrap-around« paart sich mit dreispeichigem Lenkrad, dessen Mittelspeiche wie bei vielen als Doppelspeiche ausgelegt ist. Alle Notwendigkeiten liegen in frei geformter Hülle, müssen aber den allgemeinen ergonomischen und gesetzlichen Vorgaben genügen, die zu den umfangreichsten ihrer Art am Automobil gehören. Das Markentypische geht hier wie bei den meisten Serienbeispielen wegen trendiger Gestaltung unter. Noch übertriebener wird dann ein Fahrerplatz, wenn er überfrachtet ist mit vermeintlich künstlerischen Dekors, nur weil diese auch die Karosserie heimsuchen. Damit muß die Studie BMW i8 Memphis Style von 2017 leben. Bei vielen Beispielen verwundert die Vielfalt der Anzeigen, da Ablesen eigentlich nur wenig Sinnvolles als Lösung kennt. Daß alles in klarer Designsprache machbar ist, langlebig in den Details der Mode trotzt und von selbst funktional ist, macht 2017 der Land Rover Discovery vor. Nur das kleine Detail der runden Schalter an den Lenkradspeichen will zu viel Aufmerksamkeit in ruhiger Umgebung. Aufmerksamkeit am Cockpit, von BMW seit langem bevorzugter Begriff, sieht anders aus, wenn es in die Luft geht: Beispiel dafür ist der Airbus A 320 von 2015.

British designers, who are masters at creating a club atmosphere, use this concept in their luxury marques, though the VW had the idea long ago: Its 2015 study, the Bentley EXP, revels in diamond upholstery and bewildering arrays of crystal, and gives both the driver and front-seat passenger a special »wrap-around« feeling. Its serial production sedan spin-off is equally opulent; pictured is a version that again is graced with diamond upholstery. And backseat passengers are able to sit back and enjoy a glass of champagne – but only while riding in a 2016 Bentley Mulsanne Speed. How different the ambience is when designers at Honda draft the driver's compartment of the future. The 2016 study NeuV uses the latest monitor display technology, which appears to be limitless and is a combination of everything that is necessary to control and operate a car. And then there's the challenge of figuring out how to design a steering wheel differently than usual: by simply running a ring through two guide clips. Back in the world of today the 2016 Opel Ampera sports a familiar type of dashboard: »Wrap-around« is here coupled with a three-spoke steering wheel whose middle spoke is often construed as a double spoke, as in many other models. All the necessities are located in a free-form cocoon, but have to meet the general ergonomic and legal requirements, which for cars are extremely numerous. Here, as in most series-produced automobiles, the features that are typical of a brand get lost on account of the trendy design. A driver's compartment becomes even more outrageous if it is overloaded with supposedly artistic embellishments only because the latter also infest the bodywork. The 2017 study BMW i8 Memphis Style has to live with that. What is amazing about many models is how many types of displays there are, since there is only a limited number of practical solutions for making it easy to read them. The 2017 Land Rover Discovery demonstrates that anything can be done in clear design language; long-lived, its details defy fashionable trends and this model is functional in its own right. Only the little detail of the round switches on the spokes of the steering wheel demands too much attention in an otherwise calm environment. Staying alert in the cockpit, a notion long favored by BMW, looks different when you're up in the air: A good example of this is the 2015 Airbus A 320.

363. The 2016 Opel Ampera e has the ideal »wrap-around« cockpit, echoed by the lines on the steering wheel in the center of the cockpit. (Photo: Opel.)
364. The 2015 Tesla X follows the younger generation: No more buttons, just click and swipe – hopefully accurately. (Photo: Tesla.)
365. An SUV like the 2017 Land Rover Discovery doesn't have to have a rugged design. Elegance is in. (Photo: Land Rover.)
366. The 2016 Honda NeuV rings in the era of glass displays – the wood display panel is ancient history. (Photo: Honda.)
367. The 2015 Airbus Flyer shows us what a real cockpit is. The cars are just out for show. (Photo: Airbus.)
368. In 2017 zebra stripes and an HPL pattern are good enough for the BMW study Memphis Style – but not for series production. (Photo: BMW.)

Nachwort

Automobiles Design hat es schwerer in der Beurteilung als die Daten zur Technik. PS, Beschleunigung, Hubraum, Masse, Kaskoklassen, alles ist quantifizierbar. Design ist jedoch ein wesentliches Kaufkriterium, gleichermaßen wie Zuverlässigkeit, Preis oder Verbrauch. Seit 2015 – ausgerechnet zur Zeit von Europas größter Autoschau, der IAA – wurde das Vertrauen in die Autoindustrie ramponiert durch die Abgasmanipulationen, die bei VW anfingen und bei weiteren Herstellern wie Fiat, Renault, Opel, Audi oder Mercedes vermutet wurden. Gleichzeitig werden neue Automobile präsentiert, begleitet von der Lyrik der Pressetexte, die die gestalterische Leistung der Neuen wortreich unterfüttern. Modelle der Serie werden aufgerüstet für den hybriden Antrieb, auch für den reinen Elektroantrieb. Je nach System des Antriebs wird das Design zu dessen Spiegelbild. Dieses Buch läßt die geschichtliche Entwicklung ausführlich zu Wort kommen, um darzulegen, wie die technische Entwicklung die Form des Automobils prägte, was die Außenform ebenso wie den Innenraum betrifft. Mag es vor 100 Jahren mehr Marken gegeben haben mit geringen Produktionszahlen, so ist die heutige Auswahl mit weniger Marken größer. Vergleiche der Marken und Vergleiche innerhalb einer Klasse zeigen formale Varianten mit markenunspezifischen Details. Daher beschäftigen sich viele Kapitel mit Details, die von der Designsprache eines Herstellers losgelöst sind. Das mag an ästhetischen Einzellösungen, aber auch an der verständlichen menschlichen Reaktion liegen, die auf englisch mit »keeping up with the Joneses« beschrieben wird, will sagen, was der Nachbar hat, will ich auch haben. Diese Haltung beinhaltet die Gefahr einer Inflation von Gestaltungsmaximen. Hinzu kommen die vielfach verlangten oder geförderten Ausstattungsvarianten der technischen Hilfsmittel, die das Fahren sicherer machen oder machen sollen. So geraten Grundpreise der Autos bei Vollausstattung aller Angebote der Preisliste auf fast das Doppelte. Der ökologische Ansatz verliert. Die Technik sucht Gewichtsreduzierung in Strukturverbesserungen des Materialeinsatzes. Ihr Erfolg zerbricht an den Sonderwünschen. Optimierter Verbrauch – man vergleiche diesen mit dem von vor 50 Jahren – scheitert an der Praxis jährlich steigender durchschnittlicher PS bei Neuwagen. Das Ziel einzuhaltender reduzierter Abgaswerte wird immer schwerer erreicht, der Trend zu SUVs tut ein übriges. Die Vermarktung von elektrisch angetriebenen Autos läuft dagegen schleppend und kann das kaum kompensieren, auch wenn wesentliche Kriterien wie Batteriespeicherkapazität, Reichweite, Ladezeit und Ladestationen vergleichsweise schnell verbessert werden.

So trifft es zu, daß sich die Automobilindustrie im starken Wandel befindet, sei es der Herstellungsprozeß im Vergleich zu früher, sei es die Neuorganisation des Individualverkehrssystems, wenn neben neuer Antriebstechnik die Technik des selbstfahrenden Autos die Zukunft bestimmen soll, was sicher länger dauern wird, als jetzt prognostiziert. Das mag daran liegen, daß das Bauteil Mensch in dieser Zukunft sein Verhalten langsamer ändern wird, als die machbaren technischen Möglichkeiten nutzbar werden. Das automobile Design wird Antworten darauf geben. Die Antworten heutiger Studien und Konzepte sind teilweise euphorische Übertreibungen einer möglichen Zukunft. Die junge Generation von heute und alle zukünftigen haben einen leichteren Zugang zu Automobilen, die wesentlich mehr elektronische Begleitung in der Nutzung bieten; das Smartphone war und ist dafür ein ideales Trainingsgerät. Beispiele der Innenraumauslegung speziell am Fahrerplatz belegen das. Der größte Teil des Straßenverkehrs wird von PKWs geleistet. Sie stehen im Mittelpunkt, wenn automobiles Design zum Thema wird. Zweiräder wären ein Thema, besonders dann, wenn sie verstärkt, angenähert an Karosserien, verkleidet werden. LKWs wären auch ein Thema, weil sie immer mehr nach gestalterischen Vorgaben ihre Form finden, ohne daß die Funktion vernachlässigt wird. Besonders die Fahrerplätze bei LKWs haben in Funktion und Design stark aufgeholt und sich in vielen Details der formalen Sprache der PKWs angepaßt, was allein die Lage des Lenkrads und seine leichte Bedienung zeigen. Als weiteres Thema wäre zu untersuchen, wieweit formale Entwicklung sich ökologisch orientiert. Es reicht nicht, Hutablagen aus Naturfaser zu produzieren oder gesetzliche Vorgaben für Recyclingprozesse von Altautos zu erlassen. Wie stark sich nämlich der Materialmix im Automobilbau erweitert hat, belegt die Innenraumausstattung heutiger Autos, die schon in kleinen Klassen einen Standard erreicht haben, den sich vor wenigen Jahrzehnten nur hochpreisige Autos gönnten. Die weltweite Nachfrage nach Autos hört nicht bei den Schwellenländern auf. Auch hier ist der Kundenwunsch nicht das Primitivauto, sonst hätte der indische Tata Nano mehr Erfolg. So beschränkt sich das Buch auf Regeln, Prinzipien und auf Hinweise zum besseren Verständnis des automobilen Designs. Denn die Schönheit liegt im Auge des Betrachters. Dieses aber darf geschult werden.

Afterword

Automotive design is harder to evaluate than technological data. Horsepower, acceleration, cylinder capacity, mass, bodywork types can all be quantified. But design is an important deciding factor when buying a car, as are reliability, price or fuel consumption. Since 2015 – ironically at the time of Europe's biggest auto show, the IAA – confidence in the auto industry has been shaken by the manipulations of exhaust gases that began with VW and were suspected in the case of other manufacturers such as Fiat, Renault, Opel, Audi and Mercedes. At the same time new cars are being presented, accompanied by the panegyrics of press reports that profusely describe the creative achievements of their designers. The models of a series are retrofitted with a hybrid drive, or with an all-electric drive system. the design mirrors the drive system. The present book is a detailed account of the historical development of the automobile, and describes how technical development influenced the form of the automobile, meaning its exterior form as well as its interior. While 100 years ago there may have been more brands, with low production numbers, today, with fewer brands, the selection is larger. Comparisons of brands and comparisons within a class of vehicles show formal variants with brand-specific features. That is why many chapters in this book deal with details that are examined separately from the design idiom of a manufacturer. Such details may be based on stand-alone aesthetic solutions, but also on the understandable human reaction that is described as »keeping up with the Joneses« – meaning, I want what my neighbor has. There is the danger that this attitude leads to the proliferation of principles of design. And then there is the large variety of configurations of technical devices for which there is a demand or that are promoted for the purpose of making driving safer, or supposedly safer. This almost doubles the basic price if a car is fully equipped with all available accessories. Ecological considerations lose out. Technicians seek to reduce the weight of a car by structural improvements of the material used. They cannot succeed when they have to respond to the special requests of the customers. Optimized fuel consumption – compared with consumption 50 years ago – goes overboard when faced with new cars that on the average pack more and more horsepower from year to year. It is getting harder and harder to reach the goal of observing emission standards and reducing emissions, and the trend toward SUVs does the rest. Meanwhile the marketing of electric cars is moving at a snail's pace and is hardly able to offset this trend, even if important criteria such as battery storage capacity, range, charging time and charging stations are improved comparatively quickly.

Thus the auto industry is definitely going through a profound transformation: The manufacturing process is changing in comparison to former times, and the individual transport system needs to be reorganized if in addition to a new drive technology the future is to be determined by the technology of the driverless car, which will no doubt take longer than is being predicted at the moment. The reason for this may be that in this future the human element will change its behavior more slowly than feasible technological possibilities can be utilized. Automotive design will give us the answers. The solutions provided by current studies and concepts are in part euphoric exaggerations of a potential future. The younger generation of today and all future generations have much easier access to cars, and driving involves the use of many more electronic devices; the smartphone is an ideal training tool for this, as shown by examples of vehicle interior design developed specifically for the driver's compartment. Passenger cars constitute the major part of street traffic. They take center stage when it comes to automotive design. Another important issue are two-wheelers, especially if their bodywork is reinforced, much like car bodies. Trucks are another issue, because their form is increasingly determined by design requirements, without losing sight of the function. The function and design of the driver's cabins in trucks especially show great improvement and have, in many details, adapted to the formal idiom of passenger cars, as evidenced by the position of the steering wheel and its easier handling alone. Another topic for research would be to study the extent to which formal development reflects environmental concerns. It is not enough to produce backlite shelves from natural fibers or to pass laws regulating the recycling of wrecked cars. The interior design of modern cars shows how much the mix of materials used in automobile construction has expanded: Even lower-category vehicles have now reached a standard only high-priced cars were able to achieve a few decades ago. The demand for cars is worldwide and does not stop with the emerging countries. Here too, customers do not merely want a basic car, or else the Indian Tata Nano would be more successful. The present book limits itself to rules, principles and guidelines that lead to a better understanding of automotive design. For beauty is in the eye of the beholder. But the eye of the beholder can be trained.

Glossar

Abrißkante: scharfkantige Ausbildung an der Dach- oder Heckkante zur wirbelfreien Ablösung der Luftströmung.

Advanced Design: das Austesten zukünftig möglicher Gestaltungsrichtungen von Karosserien mit Hilfe von Designstudien, z. B. auf Autosalons oder Messen.

Aerodynamik: Lehre von der Luftströmung an Körpern wie Karosserien oder Flugzeugen zur Reduzierung des Luftwiderstands oder zur Optimierung der Form.

A-Säule, B-Säule, C-Säule: Fensterpfosten, die das Dach der Karosserie tragen. A liegt zwischen Front und vorderer Seitenscheibe, B zwischen vorderem und hinterem Seitenfenster, C zwischen hinterer Seitenscheibe und Heckscheibe oder ist z. B. beim Kombi die dritte Säule vor der D-Säule als vierte.

Blister: auch als Radlippe bezeichnet, wird am Radausschnitt als ausgestellte Aufwölbung genutzt, um zum Beispiel breitere Reifen abzudecken oder den Radausschnitt zu betonen und damit zu stabilisieren.

Bombierung: kugelförmige Ausrundung der Karosseriefläche, z. B. beim Dach oder früheren Kotflügeln, wenn sie konvex oder schüsselförmig gebogen sind.

Buggy: früher leichte, dachlose Pferdekutsche, heute offene, sportlich ausgelegte Autokarosserie, meist aus Kunststoff, teilweise auf fremden Fahrgestellen und oft nur als Freizeitauto betrieben.

Cabriolimousine: Limousine mit feststehenden Seiten und voll absenkbarem Stoffdach.

Chassis: französisch für Fahrgestell, das Räder, Radaufhängung, Federung, Motor und Lenkung beinhaltet.

Clay: Modelliermasse für den Modellbau, kann leicht verarbeitet werden, ist leicht veränderbar in frischem Zustand und formstabil nach Austrocknen zur Aufnahme von Lackfolie für ein realistisches Erscheinungsbild.

Concept-Car: Studie auf bestehenden Fahrgestellen oder völlig neuer Entwurf zum Austesten, was in Zukunft machbar ist oder werden könnte und auch Vorbild für kommende Serienmodelle sein kann.

Coupé: französisch und bedeutet »abgeschnitten«. Beim Auto eine verkürzte Version der Limousine, in den meisten Fällen nur zweitürig, aber mit geschlossener Karosserie und fest montiertem Dach.

Cut-away: konkave Aushöhlung kleinerer Flächenpartien mit scharfen Rändern, beliebt bei Überbetonung von Schwellern oder zur Belebung der Flanken des Autos.

Cw-Wert: Der sogenannte Luftwiderstandsbeiwert ist nur im Windkanal ermittelbar und gibt die Qualität der luftumströmten Karosserie an. Das Produkt von Cw-Wert und Querschnittsfläche (Schattenbild der Frontansicht) ist der Luftwiderstand. Je kleiner der Wert, desto windschlüpfiger die Karosserie.

dos-à-dos, vis-à-vis: meist bei Autos der ersten Jahrzehnte die Sitzanordnung: Rücken an Rücken oder gegenüber.

Ergonomie: Lehre vom menschlichen Maßstab bei der Auslegung von Sitzpositionen, Bedienelementenanordnung, Sichtverhältnissen, Bewegungsmöglichkeiten im Wageninnern.

Facelift: Karosserieänderungen in kleinen (preiswert umsetzbaren) Schritten nach mittlerer Modelllaufzeit als Kaufanreiz, selten technisch umfangreiche Korrekturen.

Fensteröffnungslinie: der Umriß der verglasten Teile der Karosserie, teilweise verchromt, besonders wenn markentypische Linienführung traditionell gegeben ist.

Fensterwurzel: der untere Anschluß einer Säule an die Karosserie.

Gitterrohrrahmen: Leichtbau aus dünnen Rohren, der die tragende Struktur des Fahrzeugs bildet und Bleche oder Kunststoff als Verkleidung erhält. Beispiel: Mercedes SL 300.

Greenhouse: der Aufbau oberhalb der Gürtellinie. So genannt, seit der Glasanteil in den letzten Jahrzehnten größer wurde.

Gürtellinie: horizontal verlaufende, selten geschwungene umlaufende Linie zwischen dem Unterbau der Karosserie und dem Dachaufbau, direkt unter der Linie nennt man, wenn ausgewölbt, den Bereich die Schulter.

Hardtop: bei Cabrios der fest montierbare massive Dachaufsatz, etwa für den winterlichen Betrieb.

H-Punkt: sogenannter Hüftpunkt, gibt den Lagebereich an, innerhalb dessen die Hüfte in unterschiedlichen Sitzpositionen und bei unterschiedlichen Körpergrößen zu liegen kommt, um Bedienelemente (Pedale, Lenkung, Schalter) richtig zu positionieren.

Glossary

Advanced design: the testing of potential future approaches to bodywork design with the aid of design studies, e.g., at car shows or trade fairs.

Aerodynamics: the study of air movement across car or aircraft bodies with the intention of reducing air resistance or of optimizing the form.

Air scoop: a convex projection on engine hoods that is open in front and ventilates the engine compartment.

Air resistance: all the elements that exert friction against the wind, such as the cross-sectional area of the car body, poor aerodynamic design (poor drag coefficient), airflow for the interior and engine cooling system, add-ons, etc.

A-pillar, B-pillar, C-pillar: structural members that support the roof of the car body. A is located between the windshield and the frontmost side window, B between the front and rear side window, C between the rear side window and the rear windshield or (for instance, in station wagons) the third pillar in front of the fourth, or D-pillar.

Apron: located below the front and rear bumper (sometimes bearing the license number), usually a single plastic panel, often formally aerodynamically optimized.

Back-to-back, vis-à-vis: the seating arrangement in cars of the first decades.

Bathtub line: puristic pontoon-type body with an emphatic, wrap-around line, often chrome-plated, directly below the shoulder. In Germany, a feature of the 1961 NSU Prinz.

Beltline: horizontal, rarely curved, circumferential line between the undercarriage and the roof section; the area directly under the line, if curved, is called the shoulder.

Blister: bulge in the wheel cutout, used, for instance, to cover wide tires or to emphasize and thus stabilize the wheel cutout.

Buggy: formerly a light, roofless horse-drawn carriage, now an open, sporty car body, usually made of synthetic material, sometimes on the chassis and suspension of another model, and often used only as a recreational vehicle.

Camber: spherical rounding of the bodywork , e.g., in the roof or former mudguards when they are convex or bowl-shaped.

Chassis: French for the base frame of a vehicle comprising wheels, wheel suspension, springs, engine and steering.

Clay: used for building models, can be worked easily, is malleable when fresh and keeps its shape after drying, and will take lacquer foil, making for a realistic appearance.

Concept car: a study on an existing chassis or a completely new design for the purpose of testing what is or might be doable in future and might also serve as a prototype for upcoming serial models.

Convertible limousine: a type of sedan with fixed sides and a fully retractable fabric roof.

Coupé: French for »cut off«. A shortened, usually only two-door, version of the sedan, with a closed body and permanently mounted roof.

Cross-sectional area: Point at which a body has its largest silhouette area in the front view, i.e., its largest cross-section.

Cutaway: concave hollow in smaller panels with sharp edges, a favorite way of overemphasizing rocker panels or to liven up the flanks of the car.

Dashboard cowl: hollowed-out metal area between the horizontal engine hood and the windshield.

Drag coefficient (Cd): this can only be determined in a wind tunnel and indicates the quality of the car body in air flow. The product of the drag coefficient and the cross-sectional area (shadow image of the front view) is air resistance. The lower the Cd, the more streamlined the body is.

Ergonomics: the science concerned with the human scale in planning sitting positions, the layout of operating controls, visibility conditions and mobility in the car's interior.

Facelift: bodywork modifications in small, inexpensively implementable steps after an average vehicle lifetime, offered as an incentive to buy – rarely major technical corrections.

Greenhouse or glasshouse: the bodywork above the beltline, called greenhouse because in recent decades the proportion of glass has increased.

Hardtop: in convertibles this is the massive roof extension that can be firmly mounted, e.g., for winter driving.

H-point: the so-called hip point indicates the area within which the hip comes to rest in various sitting positions and with different body sizes, in order to position the controls (pedals, steering, switches) correctly.

Hutze: obere, vorgezogene Abdeckung bei Armaturen zur Verhinderung von Blendung oder Spiegelung, bei Scheinwerfern zur gezielteren Lichtlenkung. Auf Motorhauben eine vorne offene, konvexe Ausbuchtung zur Belüftung des Motorraums.

Kamm-Heck: Wunibald Kamm entwickelte einen aerodynamisch effektiven, aber kurzen Heckabschluß, der dort endet, wo der Luftwiderstand ohne lang auslaufendes Heck keinen Einfluß mehr hat.

Landaulet: kommt vom Kutschenbau und ist beim Auto dasselbe, nämlich das Absenken des hinteren Stoffdachs bei einer Limousine. Meist bei Staatskarossen, in Deutschland bekannt geworden durch den Mercedes 600 aus den 1960er Jahren.

Lastenheft: verbindliches Buch mit allen technischen wie gestalterischen Vorgaben für den Entwurf eines Autos.

Lichtkante: eine Fläche, wenn geknickt, bildet an der Knicklinie eine Lichtkante. Bekannter Vertreter Mitte der 1960er Jahre: Opel Rekord und Opel Kapitän.

Luftwiderstand: sämtliche dem Wind abträgliche Bereiche wie Querschnittsfläche der Karosserie, schlechte aerodynamische Gestaltung (schlechter Cw-Wert), Luftdurchsatz für den Innenraum und Motorkühlung, Anbauteile etc.

MPV: multi-purpose vehicle. Ein Auto meist in der One-Box-Version eines Vans wie der VW Sharan oder die ersten Renault Espace.

Off-Road: englisch für »weg von der Straße« und meint Geländewagen.

Package: Systemzeichnung für die Anordnung aller Teile des Autos innerhalb der Karosseriehülle, also Lage von Motor, Rädern, Antrieb, Sitzen, Kofferraum, Tank, Reserverad etc.

Phaeton: griechischer Sonnengott als Wagenlenker als Vorbild für den nur 14 Jahre gebauten VW desselben Namens, aber auch Karosserietyp: vom Kutschenbau abgeleitet mit Faltdach, ein bis zwei Sitzreihen, bei der letzteren Version voneinander abgeteilt, der Unterbau als Torpedo- oder Tourerkarosserie.

Pontonform: geschlossener Unterbau mit integrierten Kotflügeln, Motorhaube, Kofferraum und Scheinwerfer, am Übergang zum geschlossenen Oberbau mit betonter Schulter und umlaufender Gürtellinie.

Querschnittsfläche: wo ein Körper in der Frontansicht seine größte Schattenrißfläche, also seinen größten Querschnitt hat.

Radhaus: der Raum für das gefederte Rad, vorne vergrößert, um den Lenkeinschlag des Rades zu sichern, heute meist mit zusätzlicher Kunststoffumhüllung für vorbeugenden Schutz (Rost, Steinschlag, Verschmutzung).

Rendering: meist farbige Zeichnung, heute teilweise auch computerunterstützte Darstellung von kompletten Autos oder Details als erster Formfindungsprozeß neuer Modelle, gerne auch perspektivisch überzogen.

Roadster: offener Sportwagen mit Hilfsverdeck, kleinen Frontscheiben, oft abgespeckte Variante eines Coupés, ähnlich einem Spyder und Speedster.

Rollwiderstand: Fahrzeugmasse, Fahrbahndecke, Reifenwalkarbeit, drehende Teile des Antriebs, alles addiert sich zum Rollwiderstand.

Saloon: eine Limousine mit drittem Seitenfenster und festem Dach. In den USA sagt man dazu Sedan, in Frankreich Berline und in Italien Berlina.

Schürze: unterhalb der Stoß- oder Prallfläche (manchmal auch Träger des Kennzeichens) der Bereich an Bug und Heck, meist aus Kunststoff und aus einem Teil, oft formal aerodynamisch optimiert.

Schweller: verstärkter Karosseriebereich selbsttragender Karosserien unterhalb der Türen zwischen Vorderrad und Hinterrad, seit einiger Zeit neu entdeckte Spielfläche designerischer Phantasie.

Seitenfallung: Profillinie vom Dach bis hinunter zum Schweller in der Frontansicht, ergo Umrißlinie der Schattenrißfläche.

Sicke: langgezogene Linie, gebildet durch Flächenversatz des Bleches oder gefurchte Linie im Blech. Dient zur Gestaltung und/oder Flächenstabilisierung.

Silhouette: Wenn das Auto in der Seitenansicht als Schattenriß gedacht wird, dann ist die Silhouette die Umrißlinie der Schattenfläche.

Sitzkiste: zum Einsteigen und Hineinsetzen das Modell für den Fahrerplatz mit kompletten Ausstattungen zur ergonomischen Kontrolle und Teilen der vorderen Karosserie zur Sichtbeurteilung.

Spoiler: Luftleitflächen an der Bugschürze, am Dachende, an der unteren Heckschürze, hier meist als Diffusor ausgebildet, zur optimalen Luftumströmung der Karosserie.

Kammback: Wunibald Kamm developed an aerodynamically effective but short rear section that ends at the point where the air resistance, without a long, tapered tail end, no longer has any influence.

Landaulet: a coachbuilding term that has the same meaning when it refers to cars – a sedan car whose fabric roof can be lowered in the rear. Usually referring to state cars, the term became known in Germany thanks to the 1960s Mercedes 600.

Light edge: a surface, when kinked, forms a light edge at the kink line. Well known for having a light edge in the mid-1960s: Opel Rekord and Opel Kapitän.

MPV: multipurpose vehicle. Usually a car in the one-box version of a van like the VW Sharan or the first Renault Espace.

Off-road: an all-terrain vehicle.

Package: system drawing for the arrangement of all parts of the car within the shell of the body, i.e., the position of the engine, wheels, transmission, seats, luggage compartment, gas tank, spare wheel, etc.

Phaeton: Greek sun god, a charioteer, who inspired the eponymous VW that was manufactured for only 14 years; also, a type of car body derived from a carriage. It had a retractable roof, one to two rows of seats – separated in the latter version, and the undercarriage took the form of a torpedo or touring body.

Pontoon form: closed underbody with integrated fenders, engine hood, luggage compartment and headlights, a transition to the closed top section with emphasized shoulder and wrap-around beltline.

Rendering: a drawing, usually colored, in our days also a partly computer-generated image of complete cars or of details, the first stage in designing new models, often also three-dimensional.

Roadster: open sports car with movable top and small windshields, often a trimmed-down variant of a coupé, similar to a Spyder and Speedster.

Rolling resistance: the vehicle mass, highway pavement, tire flexing energy, rotating parts of the transmission system together create rolling resistance, or road resistance.

Saloon: a sedan with a third side window and fixed roof. While the U.S. term is *sedan*, in France it is called a *berline* and in Italy a *berlina*.

Seat box: setup for driver's compartment equipped with complete set of ergonomic controls and sections of the front bodywork for assessing visibility.

Separation edge: sharp-edged lip at the edge of the roof or the rear panel that prevents undesirable air movement across the car body.

Shoulder contour: profile line from the roof down to the rocker panel in the front view – in other words, contour of the silhouette area.

Sill or rocker panel: reinforced section of self-supporting bodywork below the doors between the front and rear wheel; for some time now, a playing field for imaginative design ideas.

Spoiler: air deflectors on the front apron, at the end of the roof, on the lower rear apron, here usually in the form of diffusers, for optimal airflow across the car body.

Strak: in German, the term for the geometric representation, by means of closely spaced section lines, of all visible interior and exterior surfaces of a car, taking into account technical and formal aesthetic requirements.

Swage line: long line formed by offset of metal panel or furrowed line in a metal panel. Used as a design feature and/or to stabilize a surface.

Targa: this term, patented by Porsche, refers to the roll bar by the fixed rear window. Between it and the windshield there are removable roof shells.

Technical specifications: an authoritative book with all technical and design-related specifications needed in order to design a car.

Three-box body: a notchback sedan with a separate hood, body (also referred to as greenhouse) and a separate luggage compartment. A two-box model is a station wagon or a hatchback (Golf), while a one-box model is a van without a separate hood and with a rear end like that of a station wagon.

Tonneau/torpedo: basic barrel-like body, sometimes with a retractable roof, which is when it is entered from the side. The torpedo variant is elongated with a continuous side profile from the radiator to the rear end and thus deeper rear seating and with side doors. Both types were built during the first automotive decades after the turn of the 19th century.

Tourer: derived from the phaeton, a closed car body with a convertible top and small doors, between 1910 and 1930. It gave its name to the later touring car.

Strak: vom Schiffbau übernommener Begriff für geglättete Flächenverläufe und ihre geometrische Verortung mittels dicht aneinanderliegender Schnittlinien.

Targa: Dieser Begriff, von Porsche geschützt, meint den Überrollbügel am festen Heckfenster. Zwischen ihm und der Frontscheibe gibt es herausnehmbare Dachschalen.

Three-Box-Karosserie (auch three-box body): eine Stufenhecklimousine mit ausgeprägter Motorhaube, dem Aufbau (auch Greenhouse) und dem ausgeprägten Kofferraum. Two-Box wäre ein Kombi oder ein Schrägheckmodell (Golf), One-Box wäre ein Van ohne ausgeprägte Motorhaube und kombigleichem Heckabschluß.

Tonneau/Torpedo: tonnenähnlicher Grundkörper, manchmal mit Faltverdeck, dann mit seitlichem Einstieg. Gestreckt mit durchgehender Seitenlinie vom Kühler bis zum Heck und somit tieferer hinterer Sitzposition und mit Seitentüren ist die Torpedo-Variante. Beides Bauarten der ersten automobilen Jahrzehnte nach der Jahrhundertwende.

Tourer: abgeleitet vom Phaeton, geschlossene Karosserie mit Faltdach und kleinen Türen, zwischen 1910 und 1930. Namensgeber für den späteren Tourenwagen.

Wannenlinie: puristische Pontonform mit betonter, umlaufender Linie, oft chrombestückt, direkt unterhalb der Schulter. Bei uns bekannt durch den NSU Prinz.

Windlauf: ausgekehlte Blechfläche zwischen horizontaler Motorhaube und Windschutzscheibe.

Windschott: bei Cabrios feststehende oder aufsteckbare Fläche aus Rahmen und dichtem Gewebe, die die Luftverwirbelung hinter den Köpfen verhindert oder zumindest abmildert.

Tubular space frame: a light-weight design made of thin tubes, it forms the bearing structure of the vehicle and is clad with metal or plastic panels. Example: Mercedes SL 300.

Wheel well: the space housing the spring-mounted wheel, larger in front in order to safeguard the steering angle of the wheel; today usually with added plastic coating for preventive protection (from rust, rocks, dirt).

Windbreak: in convertibles, a fixed or attachable panel consisting of the frame and dense fabric that prevents or at least reduces air turbulence behind the passengers' heads.

Window aperture line: the contour of the glazed sections of the bodywork, partially chrome-plated, especially when the car's brand-typical styling is traditional.

Bibliographie

Aicher, Otl, *Kritik am Auto*, München 1984.

Amado, Antonio, *Voiture minimum*, Cambridge, Mass., 2011.

Bayley, Stephen, *Harley Earl and the Dream Machine*, New York 1983.

Beattie, Jan, *Automobilbody Design*, London 1977.

Braess, Hans-Herrmann, und Ulrich Seiffert, Hrsg., *Automobildesign und Technik*, Wiesbaden 2007.

Braess, Hans-Herrmann, und Ulrich Seiffert, Hrsg., *Vieweg Handbuch Kraftfahrzeug-technik*, Wiesbaden 2013.

Buehrig, Gordon M., und William S. Jackson, *Rolling Sculptures*, Newfoundland 1975.

Caspers, Marcus, *Designing Motion*, Basel 2016.

Eckermann, Erik, Hrsg., *Auto und Karosserie*, Wiesbaden 2013.

Eckermann, Erik, *Vom Dampfwagen zum Auto*, Reinbeck bei Hamburg 1981.

Frostik, Michael, *Pininfarina – Mastercoachbuilder*, London 1977.

General Motors, *Styling – The Look of Things*, Detroit 1955.

Gurr, R. H., *Automobil Design*, Arcadia, Cal., 1955.

Hucho, Wolf-Heinrich, *Aerodynamik des Automobils*, Würzburg 1981.

Kieselbach, Ralf, *Stromlinienautos in Europa und USA*, Stuttgart 1982.

Klose, Odo, *Autoform*, Stuttgart 1984.

Krebs, Rudolf, *5 Jahrtausende Radfahrzeuge*, Berlin, Heidelberg 1994.

Lenaerts, Bart, und Linda DeMol, *Ever since I was a Young Boy I've Been Drawing Cars*, Bielefeld 2012.

Loewy, Raymond, *Hässlichkeit verkauft sich schlecht*, Düsseldorf 1953.

Margolius, Ivan, *Automobiles by Architects*, Chichester 2000.

Mende, Hans-Ulrich von, *Styling – Automobiles Design*, Stuttgart 1979.

Mende, Hans-Ulrich von, *Vorfahrt für Verführer*, Stuttgart 1991.

Mende, Hans-Ulrich von, *Porsche 911*, Frankfurt am Main 1997.

Mende, Hans-Ulrich von, *VW Golf*, Frankfurt am Main 1999.

Mende, Hans-Ulrich von, und Matthias Dietz, *Kleinwagen*, Köln 1994.

Neubauer, Hans-Otto, Hrsg., *Chronik des Automobils*, Augsburg 1997.

Norbye, Jan P., *Car Design: Structure and Architecture*, New York 1984.

Ostmann, Bernd, *Die Geschichte des Automobils*, Stuttgart 2011.

Polster, Bernd, und Phil Patton, *Autodesign international*, Köln 2010.

Schön, Christian, *Zeittafel Automobilgeschichte*, Königswinter 2011.

Seeger, Hartmut, *Basiswissen Transportation-Design*, Wiesbaden 2014.

Trapp, Thomas, *Neander*, Königswinter 2002.

Vieweg, Christof, und Harry Ruckaberle, *Mercedes-Benz Design Exterieur*, Sindelfingen 2013.

Wichmann, Hans, Hrsg., *Design Process Auto*, München 1986.

Wilson, Paul C., *Chrome Dreams*, Radnor, Pa., 1976.

Bibliography

Aicher, Otl, *Kritik am Auto*, Munich, 1984.

Amado, Antonio, *Voiture minimum*, Cambridge, Mass., 2011.

Bayley, Stephen, *Harley Earl and the Dream Machine*, New York, 1983.

Beattie, Jan, *Automobilbody Design*, London, 1977.

Braess, Hans-Herrmann, and Ulrich Seiffert, eds., *Automobildesign und Technik*, Wiesbaden, 2007.

Braess, Hans-Herrmann, and Ulrich Seiffert, eds., *Vieweg Handbuch Kraftfahrzeug-technik*, Wiesbaden, 2013.

Buehrig, Gordon M., and William S. Jackson, *Rolling Sculptures*, New Foundland, 1975.

Caspers, Marcus, *Designing Motion*, Basle, 2016.

Eckermann, Erik, ed., *Auto und Karosserie*, Wiesbaden, 2013.

Eckermann, Erik, *Vom Dampfwagen zum Auto*, Reinbeck near Hamburg, 1981.

Frostik, Michael, *Pininfarina – Master Coachbuilder*, London, 1977.

General Motors, *Styling – The Look of Things*, Detroit, 1955.

Gurr, R. H., *Automobile Design*, Arcadia, Cal., 1955.

Hucho, Wolf-Heinrich, *Aerodynamik des Automobils*, Würzburg, 1981.

Kieselbach, Ralf, *Stromlinienautos in Europa und USA*, Stuttgart, 1982.

Klose, Odo, *Autoform*, Stuttgart, 1984.

Krebs, Rudolf, *5 Jahrtausende Radfahrzeuge*, Berlin, Heidelberg, 1994.

Lenaerts, Bart, and Linda DeMol, *Ever since I Was a Young Boy I've Been Drawing Cars*, Bielefeld, 2012.

Loewy, Raymond, *Hässlichkeit verkauft sich schlecht*, Düsseldorf,1953.

Margolius, Ivan, *Automobiles by Architects*, Chichester, 2000.

Mende, Hans-Ulrich von, *Styling – Automobiles Design*, Stuttgart, 1979.

Mende, Hans-Ulrich von, *Vorfahrt für Verführer*, Stuttgart, 1991.

Mende, Hans-Ulrich von, *Porsche 911*, Frankfurt am Main, 1997.

Mende, Hans-Ulrich von, *VW Golf*, Frankfurt am Main, 1999.

Mende, Hans-Ulrich von, and Matthias Dietz, *Kleinwagen*, Cologne, 1994.

Neubauer, Hans-Otto, ed., *Chronik des Automobils*, Augsburg, 1997.

Norbye, Jan P., *Car Design: Structure and Architecture*, New York, 1984.

Ostmann, Bernd, *Die Geschichte des Automobils*, Stuttgart, 2011.

Polster, Bernd, and Phil Patton, *Autodesign international*, Cologne, 2010.

Schön, Christian, *Zeittafel Automobilgeschichte*, Königswinter, 2011.

Seeger, Hartmut, *Basiswissen Transportation-Design*, Wiesbaden, 2014.

Trapp, Thomas, *Neander*, Königswinter, 2002.

Vieweg, Christof, and Harry Ruckaberle, *Mercedes-Benz Design Exterieur*, Sindel-fingen, 2013.

Wichmann, Hans, ed., *Design Process Auto*, Munich, 1986.

Wilson, Paul C., *Chrome Dreams*, Radnor, Pa., 1976.